▶ **iMedia**

DOI: 10.1057/9781137374851.0001

Other Palgrave Pivot titles

Nicholas Pamment: **Community Reparation for Young Offenders: Perceptions, Policy and Practice**

David F. Tennant and Marlon R. Tracey: **Sovereign Debt and Credit Rating Bias**

Jefferson Walker: **King Returns to Washington: Explorations of Memory, Rhetoric, and Politics in the Martin Luther King, Jr. National Memorial**

Giovanni Barone Adesi and Nicola Carcano: **Modern Multi-Factor Analysis of Bond Portfolios: Critical Implications for Hedging and Investing**

Rilka Dragneva and Kataryna Wolczuk: **Ukraine between the EU and Russia: The Integration Challenge**

Viola Fabbrini, Massimo Guidolin and Manuela Pedio: **The Transmission Channels of Financial Shocks to Stock, Bond, and Asset-Backed Markets: An Empirical Analysis**

Timothy Wood: **Detainee Abuse During Op TELIC: 'A Few Rotten Apples'?**

Lars Klüver, Rasmus Øjvind Nielsen and Marie Louise Jørgensen (editors): **Policy-Oriented Technology Assessment Across Europe: Expanding Capacities**

Rebecca E. Lyons and Samantha J. Rayner (editors): **The Academic Book of the Future**

Ben Clements: **Surveying Christian Beliefs and Religious Debates in Post-War Britain**

Robert A. Stebbins: **Leisure and the Motive to Volunteer: Theories of Serious, Casual, and Project-Based Leisure**

Dietrich Orlow: **Socialist Reformers and the Collapse of the German Democratic Republic**

Gwendolyn Audrey Foster: **Disruptive Feminisms: Raced, Gendered, and Classed Bodies in Film**

Catherine A. Lugg: **US Public Schools and the Politics of Queer Erasure**

Olli Pyyhtinen: **More-than-Human Sociology: A New Sociological Imagination**

Jane Hemsley-Brown and Izhar Oplatka: **Higher Education Consumer Choice**

Arthur Asa Berger: **Gizmos or: The Electronic Imperative: How Digital Devices have Transformed American Character and Culture**

Antoine Vauchez: **Democratizing Europe**

Cassie Smith-Christmas: **Family Language Policy: Maintaining an Endangered Language in the Home**

Liam Magee: **Interwoven Cities**

DOI: 10.1057/9781137374851.0001

palgrave▶pivot

iMedia: The Gendering of Objects, Environments and Smart Materials

Sarah Kember
Goldsmiths, University of London, UK

DOI: 10.1057/9781137374851.0001

IMEDIA

First published 2016 by
PALGRAVE MACMILLAN

The author has asserted her right to be identified as the author of this work in accordance with the Copyright, Designs and Patents Act 1988.

Palgrave Macmillan in the UK is an imprint of Macmillan Publishers Limited, registered in England, company number 785998, of Houndmills, Basingstoke, Hampshire, RG21 6XS.

Palgrave Macmillan in the US is a division of Nature America, Inc., One New York Plaza, Suite 4500 New York, NY 10004-1562.

Palgrave Macmillan is the global academic imprint of the above companies and has companies and representatives throughout the world.

Hardback ISBN: 978-1-137-37484-4
E-PUB ISBN: 978-1-137-37486-8
E-PDF ISBN: 978-1-137-37485-1
DOI: 10.1057/9781137374851

Distribution in the UK, Europe and the rest of the world is by Palgrave Macmillan®, a division of Macmillan Publishers Limited, registered in England, company number 785998, of Houndmills, Basingstoke, Hampshire RG21 6XS.

Library of Congress Cataloging-in-Publication Data is available from the Library of Congress

A catalog record for this book is available from the Library of Congress

A catalogue record for the book is available from the British Library

Contents

DOI: 10.1057/9781137374851.0001

palgrave▸**pivot**

www.palgrave.com/pivot

Preface: A Tale of Smoke and Mirrors or Where is the *i* in *i*Media ?

Abstract: *internet, interactive, intelligent, individual...? On the one hand, there is no definitive i in imedia. Its presence may even be an illusion, a magician's trick in which branded objects are made to appear when they don't actually exist (yet) or when they are made to appear in the form of smartphones and tablets that then, with the move to transparent devices, miraculously disappear. The imedia magicians' tale is predicated on an apparently unmediated iworld that just is – that is always already transparent, filled with invisible things everywhere, at all scales and for all people, if not equally. Who sees, or who is claiming to see at scales of imperceptibility and what is at stake in this claim? While there is no definitive i in imedia, there is, on the other hand, the resurgent absent-presence of the imedia theorist and of masculine disembodied knowledge practices that come to fill the void in the turn from structure to scale and from subjects to objects as things-in-themselves. This chapter questions the in/determinability of invisible information infrastructures as subjectless objects and considers the play of critique and consensus that enables or precludes intervention. At stake is a politics of imedia predicated on a return of structure to scale as a necessary tension and figured here in the unsatisfactory and inconclusive fragments of a definition, a debate and a diagram.*

Keywords: information; infrastructures; intelligent; invisible

Kember, Sarah. *iMedia: The Gendering of Objects, Environments and Smart Materials.* Basingstoke: Palgrave Macmillan, 2016. DOI: 10.1057/9781137374851.0002.

Smoke and mirrors is a metaphor for a deceptive, fraudulent or insubstantial explanation or description. The source of the name is based on magicians' illusions, where magicians make objects appear and disappear by extending or retracting mirrors amid a distracting burst of smoke. The expression may have a connotation of virtuosity or cleverness in carrying out such a deception.

In the field of computer programming, it is used to describe a program or functionality that does not yet exist, but appears as though it does. This is often done to demonstrate what a resulting project will function/look like after the code is complete – at a trade show, for example.

(http://en.wikipedia.org/wiki/Smoke_and_mirrors)

DOI: 10.1057/9781137374851.0002

DEFINITION

A
Oct 31, 2010 5:15 AM

first of all
i need to know what does "i" stand for in many apple products such ipad, ipod, itunes...etc

is it true it is stand for "internet" or other

Thnak you

Like (0)

B
Oct 31, 2010, 12:13 PM
in response to A

Back in the day when the very first iMac was introduced, it was stated that the "i" was for "internet". Back then, it was hard to set computers up for an internet connection and the iMac was supposed make it very easy ("there's no step 3", to quote the ad from way back then).

But these days, I don't think it really stands for anything any more. People have come to associate an "iAnything" with Apple.

Why is that so important?

Like (0)

A
Oct 31, 2010 10:11 PM
in response to B

thank you for reply

but apple did not announce officaly if "i" stand for "internet" maybe another word such "intelligence"?

Like (0)

C
Oct 31, 2010, 12:21 PM
in response to A

Hi and welcome....

I was told when I bought my first iMac (1998) the "i" stood for internet

Like (0)

DOI: 10.1057/9781137374851.0002

D

Nov 1, 2010 3: 56 AM

in response to A

> but apple did not announce officaly if "i" stand for "internet"

Read the 2nd paragraph here:
http://en.wikipedia.org/wiki/iMac#History

or here:
http://my.safaribooksonline.com/0-201-70446-3/part03

or read this interview with the person who came up with the name "iMac":
http://www.cultofmac.com/20172/20172

> maybe another word such "intelligence"?

Probably not. You can continue guessing other words starting with "i" that it may be, or you can just accept what you've been told.

Like (1)

E

Jan 8, 2011 4:41 PM

in response to A

Of course you can always go for the "official" side of things like the others have posted, but a more informative approach would be that it stands for:

individual
intelligent
intuitive
inspirational
and so forth

After all, it turns out to be all of those things to those using iMachines.

Like (0)

F

Oct 8, 2011 2:57 PM

in response to A

Maybe if you look back to 1996 after Steve Jobs sold his OS "next" to apple for something like 400 million dollars, I believe. He came back as what he referred to as "iCEO" at a keynote address the "i" meaning interim CEO. Its only speculation but I would have to say it was a clever little inside joke on his behalf, seeing how since his return he took the company from the brink of falling apart to one of the most valuable tech companies in the world. so now maybe thats his way of never letting certain people forget it. Thats my guess anyways.

http://en.wikipedia.org/wiki/iCEO

Like (0)

G
Nov 29, 2011 11:10 AM
in response to A

if you recall in the 1998 Apple Back on Track Keynote, while introducing the iMac, Steve showed a slide of what the "i" stood for; and it was,

internet
individual
instruct
inform
inspire

Like (0)

H
Jun 21, 2012 6:54 AM
in response to A

I myself am 18 now and have been around Apple products: ipod, imac, iphone etc, ever since I was 8. and honestly for the last 10 years I thought the "i" stood for interactive. It makes since, "interactive phone" , "interactive macintosh" "interactive portable audio device" (I know a is not in ipod but i don't know what the o stands for) I may be right but im probably not im just throwing my oppinion out there.

Like (1)

E
Jun 21, 2012 6:58 AM
in response to H

How about injoyable

Like (1)

I
Jun 22, 2012 10:30 PM
in response to A

The answer is "internet" ... Steve Jobs declared that on May 6, 1998 Keynote.

Like (0)

DOI: 10.1057/9781137374851.0002

J
Oct 29, 2012 1:21 AM
in response to A

i in ipod stands for some thing personal, like it gives you a Sense that you own it or it make's the iPad, iPod, iMac, etc stand out in the market, How many people do you hear say "I will call them in my iPhone" instead of saying "I will call them on my phone" Apple has very cleverly branded there product name!!

Like (0)

K
Nov 4, 2012 2:24 PM
in response to E

Don't you just love those posts when they just keep going on and on? =)

Yes, the "i" May stand for any word that begins with the letter "i". so, Nobody is really sure.

Like (0)

L
Nov 4, 2012 7:10 PM
in response to A

Thanks for asking this question. It certainly prompted some fun answers especially towards the end of this thread. For my two sense may I suggest that the lower case is used for I as in "my". Works for me but no so much for earlier Macs.

Like (0)

M
Nov 4, 2012 7.20 PM

What does the "S" in "Harry S Truman" stand for?

Like (0)

https://discussions.apple.com/thread/2632371[1]

DISCUSSION/DEBATE/DRAMA

DOI: 10.1057/9781137374851.0002

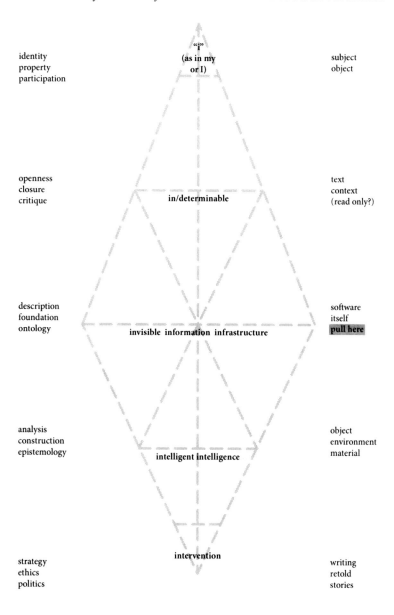

identity
property
participation

"i"
(as in my
or I)

subject
object

openness
closure
critique

in/determinable

text
context
(read only?)

description
foundation
ontology

invisible information infrastructure

software
itself
pull here

analysis
construction
epistemology

intelligent intelligence

object
environment
material

strategy
ethics
politics

intervention

writing
retold
stories

DIAGRAM

DOI: 10.1057/9781137374851.0002

DISCUSSION/DEBATE/DRAMA

There is no *i* in *i*media. Its presence is an illusion that refuses to be revealed even as branded objects like smartphones, tablets and computers are made to appear and, with the move toward transparent devices, disappear. The origin story of *i* as *i*nternet media is uncertain, contested and over-written at source. A's tentative question 'is it true it is stand for "internet" or other' is answered in the affirmative by B and C before A, acknowledging the replies, opens it up again, reaching for a perhaps more contemporary *i* for *i*ntelligent. D is having none of it. Both performing and evoking authority, D provides evidence linking *i* with *i*nternet. A's words are stripped out and highlighted as quotations, but D's scholarly bid to be right is affected by a distinct writerliness, sarcasm and staging of a conflict: 'maybe another word such as "intelligence"? Probably not. You can continue guessing other words starting with "i" that it may be, or you can just accept what you've been told.' E puts the cat among the pigeons, echoing A's ambivalence about 'the "official" side of things' and offering an open-ended list of other words starting with 'i'. Beyond the contested definition of *i*media machines, part of what unfolds here is a play of singular and plural, closed and open, fixed and indeterminable, factual and speculative knowledge. F offers an ironic reading of *i* as *i*nterim CEO, referring to Steve Jobs' heroic return to Apple that saved the company and created the possibility referred to by B, that *i* once stood for *i*nternet 'back in the day' but doesn't really stand for anything anymore. If, as B suggests, '*i*Anything' is associated with Apple, why is it so important to debate both the origin and evolution of 'i'? What is at stake beyond the anchoring or play of *i*branded words and online, community-based knowledge practices? To what extent does this mere excerpt of a discussion/debate/drama open out into the wider, if not yet ubiquitous, not yet known or knowable terrain of *i*worlds and *i*life as they are nevertheless currently in the process of being lived? The anonymized characters and participants in my utterly unrepresentative sample approach their own subtext in a spirit of serious play that is not easily reduced to either the presence or absence of critique or to the terms of the post-political. For now it is enough to suggest that the status of my sample is ambiguous.

DEFINITION

Perhaps it is G's evocation of the spectacle of smoke and mirrors (in the computer programming sense) that obviates the need for further discussion. G hones in on the 1998 trade show/demo in which Steve Jobs introduced the iMac and 'showed a slide of what the "i" stood for.' We do not need to see the slide in order to register its affect, then and now, in performing a functionality that does not yet exist. G in effect shows us – as the extended community – that the functionality of the iMac always already exceeded the *i*nternet to *i*nclude, definitively: '*i*ndividual, *i*nstruct, *i*nform, *i*nspire.' G's absent slide is doubly performative, both of the functionalities it encloses in time (past and present) and of the spectacle ('Steve showed a slide') that made them appear. *i* is then Apple's tale of smoke and mirrors, a tale that falls short of truth and that descends – tellingly, from this point onwards in the discussion/debate/drama – into what Lyotard refers to as language games (1984). The discussion is not so much ended as undone. H, in response to A, chips in with *i*nteractive but sets up the game as a game, going beyond the play of origins and opinions to something more like play itself, reoriented from contested definitions to letter substitutions. H's perhaps inadvertent substitution – 'I know a is not in ipod' – is echoed and enforced by E, who puts the cat among the pigeons again, responding directly to H and no longer to A with: 'How about *i*njoyable.' E and H obtain one 'like' each for their efforts, superseding the recognition afforded to D and reinforced, in vain, by I. As the quest for the definitive answer is derailed, the space for play itself opens up to metacommentary (K to E: 'Don't you just love those posts … ?'), new games (M to A: 'What does the "S" in Harry S Truman stand for?') and more metonymic substitutions in which, for example, i comes to stand for 'my.' L offers this substitution toward the end of a thread that also becomes a metacommentary on the thread.

DIAGRAM

i

In lieu of a mode, method or movement – the absence of which will be addressed later in the book, at some point in my second *i*Media Manifesto – I will continue rethreading the *i* in the wider word and world of *i*media that may not yet exist (although it appears as if it does) and in which more or less is at stake than the play of truth and illusion. Jodi Dean offers something close to a definition of *i*media in which *i* stands for my:

> Media, ever smaller and more integrated, are not just many-to-many, as early internet enthusiasts emphasized, but me-to-some-to-me. The rise of the consumer as producer hyped as Web 2.0 and signaled by Facebook, MySpace,

DOI: 10.1057/9781137374851.0002

and YouTube designates a shift in media such that increasing numbers of people present their own artistic work (videos, photography, music, writing), express their own views, and star in their own shows.

(Dean 2009: 4)

For Dean, the *i*media individual is isolated in both the production and consumption of their own media, engaged in self-expression rather than discussion or debate and talking predominantly to themselves as they mutter 'in a seeming dementia into a barely visible mouthpiece' (4). This almost pathologization of the mediated subject is quickly turned toward the 'academic and typing left' who, in Dean's judgment, have celebrated isolation as freedom and individual creativity and failed to counter neoliberalism or provide the logic for 'collective approaches to political, social and economic problems' (4). The notion of post-politics or of de-politicization is, for her, a manifestation of defeatism, an abdication of responsibility to market forces and ultimately a self-fulfilling prophecy. Allied to her sense of the failure and somewhat petulant irresponsibility of the left is something of a lament for the media as public sphere. The sphere of *i*media is not that of democracy but rather of the alignment between democracy and capitalism according to Dean. This new alignment or 'strange merging' is referred to as 'communicative capitalism' and characterized by the free, open, transparent and inclusive circulation of content – and by the absence of critique (20, 22). With reference to the political theory of Chantal Mouffe, Dean suggests that 'the intense circulation of content in communicative capitalism occludes the antagonism necessary for politics, multiplying antagonism into myriad minor issues and events' (24). She is referring here to lifestyle politics, ethical consumerism and perhaps even matters of mere opinion concerning the origin and evolution of the *i* in *i*media. *i*media as *my* media encompasses individual acts of expression, networks that 'turn efforts at political engagement into contributions to the circulation of content' and environments in which 'doing is reduced to talking' (32). Dean bears witness to 'the reduction of politics to communicative acts, to speaking and saying and exposing and explaining,' evincing therefore no particular faith in her own (or anyone else's) writing and underscoring precisely the sort of fatalistic and humanistic individualism and autonomy (including of markets and media from politics) that the neoliberal rationale promotes.

The problem, as Mouffe sees it, is actually not the failure of the left but the denial of antagonism in the left/right mode, not the alignment of democracy and capitalism or the elision of a public sphere but an incompatibility between liberal and democratic traditions that has been overlooked and

DOI: 10.1057/9781137374851.0002

effectively stabilized as a 'practically unchallenged' neoliberalism (2009: 5). Mouffe acknowledges Freud's sense of 'the ineradicability of antagonism' and argues against the quest for 'universal rational consensus' exemplified by the Habermasian public sphere. The idea that institutions like the media might, 'through supposedly "impartial" procedures' reconcile conflicts of interest is, she contends, misguided and potentially dangerous (2005: 3)[2]. The aspiration 'to a world where the we/they discrimination would have been overcome is based on flawed premises' including those of autonomous rather than co-constitutional or relational identity (2). Here, Mouffe draws on the Derridean notion of the constitutive outside and underlines the effectiveness of a deconstructive approach 'in grasping the antagonism inherent in all objectivity and the centrality of the us/them distinction in the constitution of political identities' (2009: 12). It is important for her to emphasize that the constitutive outside is not reducible to an us/them, friend/enemy dialectics: 'antagonism is irreducible to a simple process of dialectical reversal: the "them" is not the constitutive opposite of a concrete "us," but the symbol of what makes *any* "us" impossible' (2009: 13). The constitutive outside, understood non-dialectically, is the basis of potential antagonism that, in turn, is a condition of possibility for the political. What is more, taking account of the political by means of antagonism, or moving toward what Mouffe refers to as an agonistic politics in which enemies become adversaries, entails coming to terms with the impossibility of resolving conflict, of arriving at reconciliation or a fully democratic politics. The 'dimension of undecidability,' as Derrida emphasized, 'pervades every order' (2005: 17).

*i*ndeterminable

Beyond my random sample, how do questions of decidability and undecidability, determinability and indeterminability play out in wider discussions, debates and dramas concerning *i*media? How, in turn, do these questions impact on the conditions of possibility or impossibility concerning a politics of *i*media? It is worth noting that the notion of undecidability deconstructs the dialectics of us and them, my media and our media, corrupting any simplistic opposition or resolution of conflict but also signaling the paradox that a non-dialectical relation is both the grounds of possibility and impossibility for antagonism and agonistic politics. It is, arguably, the paradox or aporia of antagonism (as being undecidable with respect to itself) that Mouffe does not account for. Again, the constitutive outside – as

DOI: 10.1057/9781137374851.0002

my-our, individual-collective *i*media – may, simultaneously, be the grounds of possibility and impossibility for antagonism, rendering it undecidable whether an agonistic adversarial politics of *i*media can be achieved.

It is this more paradoxical or aporic sense of undecidability, characteristic of late deconstruction[3] and tied to co-constitutive or non-dialectical relations that, it seems to me, informs the wider debate on *i*media, along with a set of related concerns about in/determinability as the cause and effect of dialectical relations – the effect (or lack of it) of media on politics, the effect (or impossibility of it) of technics on time. Mark Hansen is critical of the extent to which Bernard Stiegler's philosophy of technics (a philosophy that Hansen relates to the deconstructive tradition) foregrounds its impact on the time frames of human consciousness and perception:

> ... the fact of the matter is clear: in our world today, technics does not remain indexed to human consciousness and its constitutive time frame(s) but operates at scales well outside of what humans can perceive. This ... is precisely why we must turn from perception to sensation in our effort to think the contemporary correlation of technics and life.

(Hansen 2013: 76)

This statement is notable, firstly, for its descriptive presencing of a media and technological world that is qualitatively different, if not diametrically opposed to an earlier one in which, Hansen concedes, technics were aligned with human scales. For him, 'our world today' is post-human where 'post' appears to signal 'after' and to reinstate precisely the kind of mechanistic, successional and humanistic temporality of past, present and future that he strives to displace with reference to the 'correlation of technics and life.' *i*media, for Hansen, are no longer the tangible memory objects, the photographs and books that could be subject to deconstructive critical analysis. Instead, he suggests, *i*media are the *i*nvisible *i*nformation *i*nfrastructures that require not a deconstructive, but a post-phenomenological account of sensation and experience across scales. Here, in its co-constitution in and as the *i*mperceptible *i*world of ubiquitous, *i*ntelligent or so-called smart media (in which, Hansen asserts, thinking is outsourced) the *i* of the *i*ndividual human subject is fully and finally lost, displaced, dispersed and de-presenced in a world that is, nevertheless, there.

*i*nvisible *i*nformation *i*nfrastructures

In the absence, or de-presence of epistemologically structured – gendered, raced, classed and otherwise categorized – subjects and their tangible

DOI: 10.1057/9781137374851.0002

memory objects, within a world of *i*media that nevertheless *is* (that is, ontological, or just there), how do we account for *i*nvisible *i*nformation *i*nfrastructures? Put another way, where does the post-human, post-critical, environmental, ontological/post-phenemenological, eco/egological account come from? It seems to me to come from nowhere and everywhere, from the omnipotent all-seeing (unseen) eye – or *i* meaning *I* – of the *i*media theorist. To what extent, then, are imedia I-less with respect to individual subjects and objects – the *i* as in *my* media dissolving across a sliding scale of microtemporalities and big data that only the *i* as in *I*media theorist can perceive – and how is the de- and re-presenced *i* manifest in *i*media theory? It is manifest in part through the attribution of agency to *i*nfrastructure and through subsequent debates about algorithmic *i*n/determinacy.

Ned Rossiter engages the debate about algorithmic *i*n/determinacy in a paper on Enterprise Resource Planning (ERP) and other logistical software systems that operate as forms of governance 'below the threshold of perception' (2015: 137). The degree to which 'software coupled with infrastructure determines our situation,' hinges, for Rossiter, on the degree to which that coupling is automated. Systems such as ERP appear to acquire autonomous agency and the ability to function as algorithmic forms of regulation to the extent that they are, or seem to be automated. Such systems are not 'immune from computational errors or problems associated with implementation' but are supposedly indifferent to the scale of human intervention (137). Rossiter assesses the strategy of anonymity that has 'historically been a resource for political work seeking to refuse the logic of identity and its susceptibility to apparatuses of capture' (148). He finds that it may not be feasible in the case of ERP systems 'whose real-life monitoring of movement and calculation of labor productivity require people, finance, and things to register within databases and screen interfaces as objects to manage in the interests of efficiency' (149). However, he goes on to suggest that while users must sign in to systems of work and play, these same seemingly total systems 'require a subject' in order to 'activate' a new file that generates and compiles data. The opportunity for critique and intervention is perhaps not entirely lost to the new dominion of scale? At least in the case of 'tasks not fully automated' there is scope, Rossiter suggests, for a political imaginary beyond instrumentalism and its conceptual ally – a masculinist tendency to proffer the unseen sighting of a fully systematized *i*nvisible *i*world that just is (149).

At the scale of *i*nvisible *i*nformation *i*nfrastructures, *i*media theory and *I*media theorists engage questions of *i*n/determinacy and of undecidability with respect to antagonism. In the face of (effaced by?)

automated autonomous systems[4] that extract economic value from the co-constitution of human subjects and/as data objects it is by no means certain whether or not it is possible to sustain the underlying structure of conflict or the basis of critical intervention.

intelligent intelligence

Mark Andrejevic engages this political uncertainty, inquiring into the 'fate of critique' when knowing is, or rather appears to be (appears as in a tale of smoke and mirrors) a function of databases and when knowing is outsourced to *i*ntelligent algorithms (2013: 15). He exposes the compatibility between *i*media theory and a post-critical, 'post-comprehension', post-humanly scaled *i*world that just is, and does so, in part, by problematizing the turn to emotion, affect, sentiment and sensation at the expense of symbolic logics of narrative and representation. It is these same symbolic logics that the architects of *i*worlds somewhat disingenuously – given their reliance on narrative and representation – claim to surpass (162; 18). Andrejevic's intention:

> ...is to highlight the ways in which recent theoretical developments that focus on 'things' – objects liberated from human contexts in which they are embedded and, thus, ostensibly, from human ways of knowing – align themselves with the 'post-comprehension' forms of knowledge associated with the instrumental pragmatics of the database.
>
> (Andrejevic 2013: 162)

Such pragmatics include mining, metrics, sentiment analysis and prediction which, while they might be instrumentalized for the good, encourage those who theorize and otherwise discuss, debate and dramatize *i*media to 'leave behind the challenges associated with representational forms of knowledge, narrative, and deliberation rather than working through the impasses they currently confront' (163). His argument in favor of a resuscitation of critique is also linked to the ways in which cynicism, credulity and conspiracy come to fill the void.

Recorded Future[5] is a recent addition to the predictions market. The company searches social media sites in order to predict actions and incidents and link them to people and organizations. Writing for the magazine *Wired*, Noah Shachtman[6] concentrates on the fact that both Google and the CIA have invested in the company ('no one is accusing Google of *directly* collaborating with the CIA') which 'strips from webpages the people, places and activities they mention.' It then examines spatial and temporal locations,

DOI: 10.1057/9781137374851.0002

emotion or sentiment and 'applies some artificial intelligence algorithms to tease out the connections between players.' The social media sites themselves may be open-access, but the company's software is trademarked. Satisfied with the claim that this 'goes beyond search' and sees 'invisible links' between documents, Shachtman signs off on the transparency of the deal and cites a Recorded Future spokesman who claims 'we can assemble actual real-time dossiers on people.' The unasked question is, then, which people? To what end? It is perhaps not enough for Andrejevic to signal the function creep of database monitoring when the company itself lists multiple contexts of use from defense intelligence to corporate security and when its applications are targeted at anonymous, neutralized subjects turned data objects (2013). Andrejevic also stops short of differentiating the subjects of his resuscitated critique, although he does point to inequalities between them. 'Just turning the databases back on the data miners will not be enough,' he argues, 'we need to develop resources for reviving the experience of diremption, alienation, and exploitation as such' (165). If he does not engage directly with the rich storytelling, the progressive *his*tories of search that the architects of Recorded Future trace back to Vannevar Bush,[7] he does maintain that the stories are still out there and that an intervention might be made by re-inscribing them (165).

*i*ntervention

Google's story is told in Eric Schmidt and Jared Cohen's[8] *The New Digital Age*. This offers a fully-scaled, epic account of how global connectivity – incorporating the 'five billion more people set to join the virtual world' – serves all people (if not equally) and solves conflict (2013: 13). Following a trip to Iraq, 'it became clear to us that there is a canyon dividing people who understand technology and people charged with addressing the world's toughest geopolitical issues, and no one has built a bridge' (9). Google promises to be that bridge, to become the public sphere of rational consensus that absorbs politicians – the U.S. Secretary of State is a friend, boast Schmidt and Cohen – and elides the political. As the benign custodians of this imaginary global public sphere, the authors make no secret of their investment in power:

> We believe that modern technology platforms, such as Google, Facebook, Amazon and Apple, are even more powerful than most people realize, and our future world will be profoundly altered by their adoption and successfulness in societies everywhere. These platforms constitute a true paradigm

DOI: 10.1057/9781137374851.0002

shift, akin to the invention of television, and what gives them their power is their ability to grow – specifically, the speed at which they *scale*.

(Schmidt and Cohen 2013: 10 [my emphasis])

Like a virus, Google scales up and out exponentially from databases and algorithms with their attendant pragmatics of mining, metrics, prediction and sentiment analysis to human-scaled objects like cars, phones and robots, then to environments likes homes and cities that are networked, distributed and ambiently intelligent and then on past the wearable technologies to other materials and matter on a potentially cosmic scale. Schmidt and Cohen tell a tale, a smoke and mirrors tale of technological totalitarianism, of big TED thinking[9] about 'tech-oriented solutions' to the world's social, environmental, political and economic problems (11). The subject of this tale is citizen Google – 'who will be more powerful in the future, the citizen or the state?' – who is neutralized, data-connected and effectively mute, a constitutive part of an unmediated environment that is affective, gestural, sensory and haptic (9). Citizen Google unlearns to write through 'superior voice-recognition software' that will automatically and instantly transcribe everything from emails to book chapters. This shift 'toward voice-initiated writing' is allegedly set to change the world of communicative materials. It is less likely that we will speak as we write, more likely that we will write as we speak (17). The speech-writing dialectic, in which speech is privileged in its presencing of the speaker, is fundamental to metaphysics and to its deconstructive critique. Critical intervention and invention through writing inverts and subverts the structure of all dialectics, signaling possibilities beyond patterns of presence and absence (Derrida 1997). Such possibilities are foreclosed here. Writing is returned to speech, and citizen Google has no say in his or her apparently genderless, utterly undifferentiated incorporation in, and as the environment of things that just are – or very soon will be.

However, the scalability of citizen Google, from data to passenger in a driverless car to bit part(icle) in an environment of cosmic communication (and back again) is not yet fully automated. The total technical system eliminates conflict and negates intervention but cannot conceal inequality. Its functioning therefore becomes skewed to what Schmidt and Cohen refer to as 'the upper band.' In order to demonstrate, to show and tell that 'connectivity benefits everyone' even while 'those who have none will have some' and 'those who have a lot will have more,' they invite citizen Google to imagine him or herself as 'a young urban professional living in an American city a few decades from now' before telling them a story – a literal

DOI: 10.1057/9781137374851.0002

one – about what an average morning might look like (28). Such stories are generic, a sign that writing has been at once erased and colonized, along with its strategic intent. The stories may involve a female protagonist whose name, for some reason, often begins with J, or, as is the case here, they may focus on a normative, self-regulatory, neoliberal subject, oriented toward productivity, flexibility, creativity and efficiency as increasingly feminized traits (Gill and Scharff 2013). The feminization of citizen Google, her re-ordering within incomplete scales of automation and so-called 'vanilla' system neutrality[10] occurs through *his*torical references to kitchens, chores and care and through the conjunction of women, wearables and work that connects Google with other (also ran) tech companies who write the future to be just like the past (Schmidt and Cohen 2013: 28–9). The antiquity of *i*media – incorporating gendered care robots, networked fridges and endless to-do lists in days made of translucent devices and transparent glass screens – is specific to the 1950s, an era in which the correlation of technics and time distilled domestic science, productive reproductivity and labor efficiencies in feminine forms and in feminized times and spaces. The story that stretches from Monsanto's to Microsoft's home of the future and that writes citizen Google into existence need not be one in which she has no say, even if, with Carol Stabile[11] and others, we acknowledge the alignment, the non-dialectical relation between feminism and neoliberalism.

The critical value of a retold (again) story of *i*media is not enclosed but is perhaps indicated in the discussion, definition and diagram that I presented at the beginning of my now quite extensive – for a short book – preface. The diagram invites Google's (Apple's, Microsoft's, Corning's) subject to pull at the threads that make up and can remake a patchworked,[12] collaborative tale of smoke and mirrors. Pulling at, and re-threading *i*'s can collapse the story's structure and yank *i*nvisible *i*nformation *i*nfrastructures from its core. A good tug could connect *i* as in *my* (or *I*)media with the possibility of intervention. Rethreading occurs across structure and scale. It is a post-human activity in which post- is not after and software subjects still know how to write. The one-act, first-page drama in which the quest for a definitive *i* is less opposed, more undone (the tale of smoke and mirrors is contingent, for its telling and retelling on an incommensurability with truth) is a manifestation of speech-writing. It is a deconstructive act that predicates antagonism as that which is undecidable with respect to itself and that is (as it were, in itself) neither political nor otherwise. Insofar as its status remains ambiguous, the possibilities for a wider debate remain open.

DOI: 10.1057/9781137374851.0002

Notes

1   The names of participants in this discussion have been removed. Links included by original contributors anonymized as D and F have been removed by the publisher.

2   Potentially dangerous in that the attempt to elide conflict leads to the possibility of more conflict. Mouffe argues that: 'the consensual approach, instead of creating the conditions for a reconciled society, leads to the emergence of antagonisms that an agonistic perspective, by providing those conflicts with a legitimate form of expression, would have managed to avoid' (2005: 4).

3   Jack Reynolds identifies two forms of undecidability in Derrida's work. In its most familiar form it is 'one of Derrida's most important attempts to trouble dualisms.' Subsequently, Derrida plays with possible-impossible aporias so that 'it becomes undecidable whether genuine giving, for example, is either a possible or impossible ideal.' http://www.iep.utm.edu/derrida/

4   The critique of autonomy as a humanistic fantasy and form of splitting and projection is of course, well established in feminist science and technology studies (Kember 2003; Suchman 2007; Hayles 1999; Adam 1998)

5   https://www.recordedfuture.com/

6   http://www.wired.com/2010/07/exclusive-google-cia/

7   Widely credited as one of the original architects or 'fathers' of the information society due to his work on the memex or memory machine in 'As We May Think', *The Atlantic*, July 1945

8   Executive Chairman and Director of Google Ideas respectively.

9   I recommend Umair Haque's critique of TED thinking, 'Let's Save Great Ideas from the Ideas Industry', *Harvard Business Review*, March 6, 2013 https://hbr.org/2013/03/lets-save-great-ideas-from-the

10   See Rossiter (2015), vanilla systems like Enterprise Resource Planning (ERP) are regarded as neutral in relation to, for example, surveillance systems such as Face Recognition Technology (FRT).

11   Carol A. Stabile 'Magic Vaginas, The End of Men, and Working Like a Dog', paper presented at Goldsmiths, University of London Annual Gender Event: Feminism and Intimacy in Cold, Neoliberal Times (2013). I'm grateful to Carol for allowing me to cite the pre-publication version.

12   My threading metaphor is indebted to the wonderful work of Kristina Lindström and Åsa Ståhl for whom it is key to a feminist interventionism in media, design and technology (2014). Viewed at an angle (in other words, from somewhere, from a perspective) my diagram is is not a diamond made of glass but a bit of fabric, a patch, to be reworked.

DOI: 10.1057/9781137374851.0002

1

*i*ntroduction: Objects, Environments and Materials

Abstract: *The imedia industry is making up stories about the world that theorists of imedia (or what is left of what Jodi Dean refers to as the 'academic and typing left' [2009: 4]) are in danger of being complicit with. How so? I argue that there is currently too much emphasis on what is, or is to come and on the failure or irrelevance of critique as something negative, uncreative and unworldly. In place of critique, there is a tendency to affirm and celebrate the existence of objects, environments and materials in themselves; to elide writing and other forms of mediation and to engage in disembodied knowledge practices. Disembodied knowledge practices tend toward both scientism (including physics envy) and masculinism. It transpires that there is no 'we' in t(he)ory[1]. Where there is some degree of convergence, some compatibility between masculinist and feminist thinking about objects, environments and materials, there is also divergence, conflict and the possible opening toward a politics of imedia. For me, this possibility hinges on the antagonism between a Harmanesque speculative realism and Harawayesque speculative fabulation/fiction/feminism and also on the non-dialectical relation of structure and scale, objects and relations, epistemology and ontology (2010; 2011).*

Keywords: masculinism; materiality; materials; object oriented ontology; scientism; sf; speculative realism

Kember, Sarah. *iMedia: The Gendering of Objects, Environments and Smart Materials*. Basingstoke: Palgrave Macmillan, 2016. DOI: 10.1057/9781137374851.0003.

What's the story – or, when is a list not not a story?

smartphone, smart watch, smart bra, translucent glass, driverless car, drone, robot, avatar, algorithm, database, data, data mining, BIG data, augmented reality mirror, smart toilet, speech recognition technology, face recognition technology, sensor, actuator, networked distributed intelligent computing, artificial intelligence, ambient intelligence, ambient media, infrastructure, smart environments, autonomous weapons, Willow Glass,™ Gorilla Glass, smart home, smart city, Siri, Google Glass (RIP), *A Day Made of Glass,* hologram, motion detector ...

Your apartment is an electronic orchestra, and you are the conductor. With simple flicks of the wrist and spoken instructions, you can control temperature, humidity, ambient music and lighting. You are able to skim through the day's news on translucent screens while a freshly cleaned suit is retrieved from your automated closet because your calendar indicates an important meeting today. You head to the kitchen for breakfast and the translucent news display follows, as a projected hologram hovering just in front of you, using motion detection, as you walk down the hallway. You grab a mug of coffee and a fresh pastry, cooked to perfection in your humidity-controlled oven – and skim new emails on a holographic 'tablet' projected in front of you. Your central computer system suggests a list of chores your house-keeping robots should tackle today, all of which you approve. It further suggests that, since your coffee supply is projected to run out next Wednesday, you consider purchasing a certain larger-size container that it noticed currently on sale online. Alternatively, it offers a few recent reviews of other coffee blends your friends enjoy.

(Schmidt and Cohen 2013: 29)

What is a list? (What) is it of? (What) does it do? To what extent does it describe a world or perform it? Is it a genre, a kind of writing or, as Ian Bogost claims, a 'flat ontology' of multi-scalar things that are (2012: 18). For Bogost, writing is at odds with worldliness: the list; the inventory; the catalogue evincing 'the abandonment of anthropocentric narrative coherence in favour of worldly detail' (41). For him, a song or a photograph that lists, does not mediate, and a book of lists, by extension would constitute an 'ontographic machine' (52). For me, listing is a literary tradition, one that reaches an apotheosis in Georges Perec's *Life A User's Manual* (2003). In as far as this gestures beyond narrative and representation, it does so, and can only do so deconstructively, from within. Perec's list does anti-literary work, performing an act of narrative creation and destruction. Bogost's list pertains to be of the world, to

DOI: 10.1057/9781137374851.0003

pull off the metaphysical trick of realism without representation. It is not authored, but existing: not language, but things. What about my list? My list is authored and *i*ncomplete. It picks up on the current cataloguing of smart objects, environments and materials that, like driverless cars and smart toilets, just are/coming soon. My list intersects an industry story set in the very near future and brought to us on translucent screens. It fills the gaps in a story that relies on sensors, speech-recognition and home ambient intelligence systems whose existence is not yet complete.

The story Eric Schmidt and Jared Cohen – the executive chairman and director of ideas at Google – tell, of a day in the work-life of a neutralized citizen, is casually sexed (breezing by the *his*toried, feminine connotations of kitchens, chores, consumption and coffee mornings), while Bogost's disavowal of writing and storytelling (including his own) is casually sexist, concerned with listing as metaphysical man-talk[2] and content with the world and all of its inequities just as it is. Bogost tells the story of what happened when his conference website or 'image toy' generated a sexist image:

> The trouble started when…one of the symposium speakers, related to me that a (female) colleague had showed the site to her (female) dean – at a women's college, no less. The image that apparently popped up was a woman in a bunny suit.
>
> (Bogost 2012: 98)

His language is disparaging, dismissive and jokey. His use of parentheses indicates with whom the joke is supposed to be shared. There is some irony in the telling – 'a sexist "toy" on a website about an ontology conference organized by and featuring 89 per cent white men' – as well as a grudging account of how he eventually conceded in destroying 'the gadget's ontographical power' by rewriting its algorithm. The alteration 'solved the problem' and stayed the trouble, but:

> The change also risks excluding a whole category of units from the realm of being! Are women or girls or sexiness to have no ontological place alongside chipmunks, lighthouses and galoshes?
>
> (Bogost 2012: 99)

Women, girls and sexiness alongside chipmunks, lighthouses and galoshes? So much for the neutrality, the in-significance of lists and so much for 'flat' ontology.

DOI: 10.1057/9781137374851.0003

Objects and OOPs theory

Bogost aspires to be among a loose association of male metaphysicians who expound Object Oriented Ontology (OOO), also known as Object Oriented Philosophy (OOP). What is Object Oriented Philosophy – or as I shall refer to it, OOPs theory – and in what sense is it[3] masculine as well as male-dominated? Claiming to relinquish the philosophical dichotomies of naturalism and relativism, realism and idealism, Bogost's OOPs theory flips the hierarchical binary of epistemology and ontology, humanism and post-humanism on the basis that (paraphrasing Graham Harman paraphrasing Martin Heidegger): 'objects do not relate merely through human use but through any use, including all relations between one object and any other' (5). He is more invested in objects as things-in-themselves than in relations that might reduce objects to processes.[4] All things count as objects and 'all things equally exist, yet they do not exist equally' (11). Bogost, not unlike Schmidt and Cohen, is precluded from attending to structural inequality by virtue of his emphasis on scale and by his association of politics and critique with solipsistic, humanistic knowledge or epistemology as 'the sieve of humanity' (3). While, with some justification, he criticizes the back-door humanism of post-humanist perspectives on ecology (I will return to the human-ist endgames of the anthropocene shortly), he disavows the inevitable anthropomorphism of his own so-called 'alien' phenomenology that asks not 'what does my body know of Photography' – as Roland Barthes did – but rather 'what's it like to be a computer?' (Barthes 1982: 9; Bogost 2012: 9). In order to answer the question, he turns to the ontography of listing, the fantasy of things that write themselves, and it is here, in his recourse to an *i*-less mediation replete with objects but devoid of subjects that the absent *i*media theorist surfaces as the return of the repressed – the masculine subject of disembodied knowledge. Bogost's alien phenomenology is both a calling card and a point of departure from Harman's object-oriented ontology, recognizing, as it does, the inevitable dialectic of object and subject as well as of material and immaterial things and becoming fatally attracted to a 'what's it like' approach to all things at all scales, including, especially, inanimate 'non-things, properties and symbols' (23).

Harman is the founding father and principal *hi*storiographer of Object Oriented Philosophy. His OOPs theory is an egocentric account of how he came to account for objects 'in their own right' (2010: 14). The objects

DOI: 10.1057/9781137374851.0003

that exist as ethical imperatives, 'commanding one another by way of the reality of their forms,' are the apples of his unseen, all-seeing *I* (21). Remaining distinct from them and enchanted by them, he is free to implicate himself in a *his*tory of great object oriented philosophers. It is to the philosophers, rather than their ethical objects that he relates. It is to them that he holds himself accountable, above any other animate or inanimate things. Like Haraway, he draws on Alfred North Whitehead's philosophy of 'actual entities' (that are relational, processual), but unlike her, he does so without reference to Michel Foucault's critique of Enlightenment epistemology or to structures of power/knowledge that she figures as hierarchical, highly differentiated relations between knowing subjects and co-constitutive objects. Where Haraway draws an epistem-ontology from the diffraction of Whitehead and Foucault,[5] Harman pursues an ontology of entities that affect other entities, inspired by Whitehead's physics-inspired metaphysics (32). He wants to speak – although he does not – of 'wood itself,' not wood as a symbol, metaphor, trope or literary figure. The linguistic, anti-realist turn against which he turns prevented philosophers from accessing inanimate worlds. These were left to the physicists, but now, Harman declares, it's time to get them back:

> When rocks collide with wood, when fire melts glass, when cosmic rays cause protons to disintegrate, we are asked to leave all of this to the physicists alone. Philosophy has gradually renounced its claim to have anything to do with the world itself. Fixated on the perilous leap between subject and object, it tells us nothing about the chasm that separates tree from root or ligament from bone. Forfeiting all comment on the realm of objects, it sets itself up as a master of a single gap between self and world...
>
> (Harman 2010: 94)

Harman's dismissal of a philosophy that has been minding the wrong gap – the epistemological, not the ontological gap – is necessarily a counter-reactionary one. By aligning deconstruction with anti-realism, he obliges himself, as a realist, to elide it, to decry it as dogma and to reinforce the very dialectic – of ideal/real, self/world – that his philosophy cannot effectively critique. His philosophy comes to be an OOPs theory not only by means of his own all-too-present absence as a philosophical subject, but in the elision of critique in favor of the endless circularity of philosophical in-action and counter-reaction. Even internally, within the trade-off between object, and relations oriented ontologies (Whitehead is ultimately too relational for Harman), there is only an

DOI: 10.1057/9781137374851.0003

oscillation between poles – such as things and processes – and nowhere near enough momentum to reach out to the world that this philosophy claims to reach out to, no attempt to speak of wood itself rather than how philosophy might be reoriented to doing so.

Environments as 'derangements of scale'[6]

The fetishization of objects serves to align markets and metaphysics. It makes *i*media OOPs theory compatible, if not necessarily complicit with, industry goals oriented to novelty and innovation – iPhones[n] – rather than invention and intervention. Intervention is precluded in the shift from knowledge to things, and environments of things-in-themselves. Environments of *i*media incorporate and exceed human subjects, operating, it would seem, at scales way below and above human perception and humanistic narrative and representational frameworks. Tom Cohen refers to these frameworks as part of our 'homeland security,' a humanistic defense mechanism increasingly threatened by environmental forces that are much bigger (and smaller) than us (2012: 18). I am interested in how *i*media environmentalism – oriented toward the ubiquity of objects and/or relations – produces what Timothy Clarke calls 'derangements of scale,' derangements in which the *i*media subject is lost and found, chaos leads to order (again) and the conditions of possibility for a new foundationalism emerge (2012: 148).

I am not opposed to *i*media environmentalism per se. With Joanna Zylinska, I have argued that it has the potential to open out from the limited dualism of technology and use, technicism and humanism and reorient our thinking from just being-in, to becoming-with the technological world (2012). However, our vitalist account of *i*mediation is a critical one, concerned with the boundary cuts between co-constituting processes and the instrumentalisation, and potential re-instrumentalisation of forms. It is in boundary work – as critical-creative decision-making – that a politics and ethics of *i*media are enacted. The absence of boundary work allows politics to default to the mainstream, to what *is* (or is soon, inevitably, to be) and produces derangements of scale in *i*media theories that are very small and/or very big. Scales slide in the absence of epistemological structures, the boundary cuts of particles and people, black and white, wood and trees that are too easily and conveniently conceded to, uncontested, made uncontestable and consigned to the philosophical past so

DOI: 10.1057/9781137374851.0003

that we can move on to better things-in-themselves. These 'things' slide all the way from code to cosmos and back again, carried, as they increasingly are, by science, not the human/ities, by computer science, physics, chemistry, biology and ecology. In big-small *i*media theory, in which the cut of the post-human is already done and dusted, it is no longer all about the economy (stupid). It's all about the ecology. It's about an infinity of *i*nvisible *i*nformation *i*nfrastructures or *i*nterstitial grey media evolving a world that is fundamentally with/out us – beneath and beyond.

If, as I've been suggesting, we ask who writes this *i*-less story of *i*media and from where (the new gods of *i*media are everywhere and nowhere. They alone can perceive the *i*mperceptible. They alone have the power of pure revelation) we may also inquire, to what end? There are, as I've argued elsewhere,[7] endgames afoot. The gods have our (human) salvation and damnation in mind. The salvationists are the big *his*torians of *i*media, seeking to secure the fate of techno-humanity before the entropic end of the anthropocene. The aim here, as the biggest of all big *his*torians, David Christian puts it is 'to pass the world on in good shape to our heirs' and find 'more sustainable ways of living' (2011: 471, 475). The damnationists, on the other hand, revel in the prospect of extinction, of an ecologically assured wipeout that would enable every smaller than human thing to begin again (Cohen 2012). It is notable that *his*torians of *i*media, big and small and invariably god-like, are working with scientific concepts of complexity and chaos theory as if they had never been worked with before. Twenty years ago, Katherine Hayles, Vivian Sobchack and Isabelle Stengers examined such phenomena with relation to the culture and politics of postmodernism (1991; 1990; 1989). Specifically, they marked a roundabout return to humanistic narratives, unified theories and representational frameworks from within their own apparent dissolution. For Sobchack, chaos was an alternate route to order and control manifest through iterative patterns of fractal geometry and self-similarity across scale. She regarded it as a universal, totalizing worldview even as it romanticized itself as the emergence of a brand new science – or as the re-emergence of science itself. Chaos and complexity were always allied in their evocation of systems with magical, supernatural properties, including the spontaneous self-organization of things. They pulled rabbits out of entropic hats and, through simple analogy, returned technological systems to their origins in nature. Every complex system, not just the weather, was subject to the butterfly effect of sensitive dependence on initial conditions. If these conditions could

DOI: 10.1057/9781137374851.0003

not be predicted or known in the classic, positivistic sense, if they could not be quantified, measured and controlled, then what could be was the pattern they eventually gave rise to. It is worth observing that what is not being evoked again is the strange attractor, the infinite, always drawn and never static figure of eight that depicted rather than described the boundary between objects within chaotic, complex (deranged) systems[8].

Smart materials are not things-in-themselves

> HAMM: Nature has forgotten us
> CLOV: There's no more nature
> HAMM: No more nature! You exaggerate

> (Beckett 1958: 16)

The absence of boundary work, the elision of structure in scale, epistemology for ontology and relations for objects signal the conditions of possibility for the re-emergence of foundationalism (Brown 2005). Citing Wendy Brown, Dennis Bruining describes foundationalism as a residual belief, or at least a willing suspension of disbelief in presence, foundation and truth after postmodernity (2013). Characterized as the new materialism or as the material and ontological turn, the principle object of foundationalism is not, currently, nature, but matter itself. Matter has become the material manifestation of life it self, the outcome of inverting a false dichotomy – and an old one – that does nothing to deconstruct it. As I've already suggested, the turn to object foundationalism claims to eschew old dialectics such as ideal/real only to end up reinforcing them. Matter is the new real. The anthropologist Tim Ingold seeks to move beyond the Aristotelian distinction of form (morphe) and matter (hyle) which is reproduced, he argues, in current attempts 'to restore the balance between its terms' (2010: 2). Ingold aims to replace the dialectical structure of form and matter 'with an ontology that assigns primacy to processes of formation as against their final products, and to flows and transformations of materials as against states of matter' (3). Operating through a Bergsonian philosophy of process that privileges movement, flow, time and creative evolution as synonyms for life itself, Ingold corrects Bergson's life/matter, time/space distinction by creating his own divisions between objects and things, agency and life, materiality and materials:

DOI: 10.1057/9781137374851.0003

I shall argue that the current emphasis, in much of the literature, on material agency is a consequence of the reduction of things to objects and of their consequent 'falling out' from the processes of life. Indeed, the more that theorists have to say about agency, the less they seem to have to say about life; I would like to put this emphasis in reverse.

(Ingold 2010: 3)

A non-anthropocentric focus on life processes demands a shift in emphasis, Ingold argues, away from the abstract concepts of materiality and material agency to materials as things-in-themselves. We would learn more, he suggests, 'by engaging directly with the materials themselves,' and following what happens to them 'as they circulate, mix with one another, solidify and dissolve in the formation of more or less enduring things' (2011: 16). To the extent that Ingold's is an *i*-less theory that reaches beneath, below and beyond subject-object driven agency, it does attempt to address the problem of communication and mediation, the relation of knowledge and life that Bergson sought to reconcile. Following Bergson, Ingold, like the political theorist Jane Bennett, seeks a knowledge formed through direct engagement with life, through contact and openness rather than an intellectual mastery and control that Bennett associates with the idea of disenchantment (2001). Her argument in favor of a re-enchantment with the world has endeared her to object foundationalists and masculine metaphysicians (she is one of very few women cited in Bogost's *Alien Phenomenology*) while she is also one of the cornerstones of material feminism – 'a strikingly affirmative form of feminist theorising' as Maureen McNeil puts it (2010: 427). While Bennett's political theory might fall short of a political theory as Mouffe characterizes it because of its emphasis on an *ethics* of openness,[9] it is certainly strikingly affirmative. For Bennett, 'enchantment is a feeling of being connected in an affirmative way to existence' (156). It is about 'saying "yes" to the world,' loving life 'before you can care about anything' (4). Attuned not only to the vibrancy of matter but to the reparation of loss, Bennett's theory bridges OOPs theory and two kinds, two contemporary manifestations of dialectical feminism – those associated with the material and reparative turns respectively:[10]

The depiction of nature and culture as orders no longer capable of inspiring deep attachment inflects the self as a creature of loss and thus discourages discernment of the marvelous vitality of bodies human and nonhuman, natural and artifactual. While I agree that there are plenty of aspects of

DOI: 10.1057/9781137374851.0003

contemporary life that fit the disenchantment story, I also think there is enough evidence of everyday enchantment to warrant the telling of an alter-tale.

(Bennett 2001: 4)

Bennett's alter-tale is reactionary and foundationalist. It seeks to establish the grounds – an openness to the presence of matter that is in itself ethical and ontological – for a more positive, affirmative politics. This grounding in turn substitutes cynicism, paranoia and critique for a more creative and generous outlook, albeit on a cynical, paranoid and critical world. For Bennett, we can and should[11] continue to feel 'wonder and surprise' even though enchantment and mystification double-time as marketing and advertising strategies, as key to 'a cynical world of business as usual' as well as processes of normalization and discrimination (8, 113).

I share McNeil's concern about the abandonment of critique in what I'm referring to as dialectical feminism. For her, it is symptomatic of a turn away from second-wave feminism and its analysis of patriarchal canons, the absence of women from fields of inquiry and male-oriented knowledge (2010: 428). The compatibility between dialectical feminism in the material mode and masculine, male-dominated metaphysics is, arguably, one outcome of this turn away. The abandonment of critique also characterizes the reparative mode that McNeil traces through the work of Elizabeth Grosz and back to Eve Kosofsky Sedgwick. Sedgwick argued for a disavowal of a 'hermeneutics of suspicion' in favor of a more conciliatory and reparative reading. She 'proposes an approach to reading and intellectual encounter which is open, emotional and positive,' a mode that 'contrasts markedly' with the more distancing and rationalist aspects of critique (432). If, for McNeil, the abandonment of critique as high-minded detachment, or as a limited strategy that failed to yield social and political change is dangerous – 'because we are still living in a world that is far from ideal' – for me, it is also only possible by adopting a limited critical strategy or a dialectical form of critique (433). Ultimately, the disavowal of critique is always 'somewhat disingenuous' (433). It is marked, as I've suggested, by limited strategies of philosophical in-action and counter-reaction, by the substitution of this for that and, in the context of contemporary feminism in the material mode by a tendency to reassert rather than reinvent nature.

DOI: 10.1057/9781137374851.0003

Towards a non-dialectical feminist perspective

Haraway's work on the reinvention of nature operates through deconstructive figures (the cyborg, modest witness, companion species), provisional concepts (material-semiotic, natureculture, FemaleMan) and retold stories that 'reverse and displace the hierarchical dualisms of naturalized entities' such as subjects and objects, people and particles, wood and trees (1991: 175). It goes beyond strategies of substitution in order to break down dialectical structures and its principal mode is writing. Writing, for Haraway, is un-aligned and undecidable with respect to binaries. It is not critique any more than it is creativity. It is not humanistic any more than it is animalistic, or reducible to words any more than it is a means of presencing worlds. Writing is a mode of reinventing the world without having to affirm or deny it. It is, I would suggest, writing more than critique that is abandoned in dialectical feminism[12] and with it goes the strategy of reinvention that Haraway currently figures in terms of sf:

> Sf is that potent material semiotic sign for the riches of speculative fabulation, speculative feminism, science fiction, science fact, science fantasy – and, I suggest, string figures. In looping threads and relays of patterning, this sf practice is a model for worlding. Sf must also mean 'so far,' opening up what is yet-to-come in protean times pasts, presents, and futures.
>
> (Haraway 2012: 4)

Terrapolis is Haraway's word for her (not a) reinvented sf world. In her account of sf, she lists all the things that Terrapolis is, without claiming to substitute authorship and narrative for worldly detail. Her concept of sf signals beyond the circularity of anti/realism and is not the opposite but rather the constitutive outside to Harman's OOPs based speculative realism – or sr. Sf and sr co-exist in an antagonistic relation in which co-existence in pasts, presents and futures is at stake. Melanie Sehgal's reading of Haraway in her relation to Alfred North Whitehead and Karen Barad suggests to me that the difference, or the key difference between sf and sr might lie in Haraway's attempt to diffract the world and Harman's claim to reflect it just as it is[13] (2014). Diffraction is a phenomenon of interference 'generated by the encounter of waves, be it light, sound or water and, within quantum physics, of matter itself. Such a superposition of waves produces a diffraction or interference pattern that records, i.e. incorporates the trajectory of the waves' (188). Diffraction is also a

DOI: 10.1057/9781137374851.0003

metaphor for, as well as a method of, engaging with a world that includes other worlders, other storytellers. Sehgal shows how, through her diffractive relations, Haraway is invested not just in epistemological interference but also in ethico-epistem-ontological interference across scales and states of existence. Crucially, this interference doesn't just happen among things-in-themselves. It is not a pure phenomenon that can be dispassionately observed and reflected. The pattern it generates and the trajectory it describes do not write themselves. Diffractive phenomena are situated, viewed and enacted from somewhere.

In its antagonistic relation to sr, one in which the degree of interference might be said to be heightened leading to a clashing of waves, sf posits a non-dialectics of structure and scale, epistemology and ontology, objects and relations. In her list-story, Haraway's Terrapolis 'is abstract and concrete.' It is 'a niche space for multispecies becoming-with' and it is 'multi-scalar' storytelling (2012: 4–5). Sf offers me a guide, a way of dealing with and even diffracting the derangements of scale in *i*media theories of objects, environments and materials in themselves.

Notes

1   I have already argued, in the preface, that there is no 'i' in *i*media. It would appear that *i*media environments – material and discursive – are post-human if not post-masculinist. At a recent conference, a male colleague spoke, I felt, for many others when he announced, enthusiastically, and from his position as conference organizer, chair and speaker that in the era of i-less media, there was 'no more man'. No more man, say the men. Surely they are protesting too much?

2   Listing is a key manifestation of what Bogost calls 'ontography.' Derived from a work of fiction, this concept is then traced through a number of thinkers – all male – associated with, or related to object oriented philosophy (2012).

3   No homogeneity implied. I read two specific accounts of a non-unified field that nevertheless address the re/turn to metaphysics from anti-metaphysics.

4   Here, Bogost distinguishes his 'alien phenomenology' from process philosophy and from actor network theory, both of which are, for him, too invested in dynamics (2012: 6).

5   Or indeed, an ethico-epistem-ontology from the diffraction of them and Karen Barad (Sehgal 2014).

6   See Timothy Clarke, 'Scale' in Tom Cohen's *Telemorphosis* (2012: 148).

DOI: 10.1057/9781137374851.0003

7   See Sarah Kember and Joanna Zylinska, 'Media Always and Everywhere. A Cosmic Approach', in Ulrik Ekman (ed) *Ubiquitous Computing, Complexity and Culture*, New York and London: Routledge (forthcoming).

8   The Lorenz, or strange attractor. A Wikimedia Commons featured picture http://en.wikipedia.org/wiki/File:Lorenz_attractor_yb.svg#filelinks

9   Mouffe distinguishes between ethics and politics in a way that might itself be too stark, too oppositional (2013).

10 See *i*Media Manifesto Part 2 for a brief discussion of the reparative turn.

11 One characteristic of dialecticism in Bennett's work as in others is a certain moral undertone that is compatible with the turn to truth, rightness and foundation in nature.

12 Clare Hemmings also implies this in 'The materials of reparation' (2014).

13 Even though the world as it is is not altogether outside of the human mind for Harman, hence his use of the word 'speculation' (2010: 2).

DOI: 10.1057/9781137374851.0003

2

*i*Media Manifesto Part I: Remember Cinderella: Glass as a Fantasy Figure of Feminine and Feminized Labor

Abstract: *Glass is the ubiquitous imaterial of the day. It has come to incorporate the properties of plastic and is promoted as an intelligent skin, covering and protecting the data subjects, objects and environments of imedia by making everything (equally) clear, open and transparent. Glass is a great leveler. It is tied to an ongoing history of democratization and transformation, of manufacturing and magic. It works harder than ever to make everything simply appear (equal) but is itself indelibly marked as an imedium, an in-visible creator of iworlds. Glass is therefore ambiguous. It is a liquid-solid fantasy figure, making up once-upon-a-time histories of a world that was, is, or very soon will be. If the world that glass made (up) once featured Cinderella and her slipper, the world that glass is currently making features as Cinderella and her slipper. Glass is becoming its own fantasy figure of feminine and feminized labor. No longer merely decorative, domestic or functional, glass works – unloved and un-seen – toward its own final transformation into the bodies it protects, the foot that it fits. Remembering Cinderella means recognizing glass as an imaterial fantasy figure and maker up of worlds that have always been, and continue to be, contested.*

Keywords: Cinderella; feminized labor; opacity; smart glass; transparency

Kember, Sarah. *iMedia: The Gendering of Objects, Environments and Smart Materials*. Basingstoke: Palgrave Macmillan, 2016. DOI: 10.1057/9781137374851.0004.

DOI: 10.1057/9781137374851.0004

Glass is the ubiquitous *i*material of the day

Materials manufacturer Corning put together a futurist video last month called 'A Day Made of Glass,' which has spiraled into stratospheric popularity on YouTube. The premise of the video is that we're about to live in an era of ubiquitous touch screens (made with glass) and smart windows (made of glass) as well as appliances like stoves which are also made with glass.

What's striking about the video is mostly how we see the touch screens working, and the way the mobile devices seamlessly network with household appliances, TVs, and bendable flat screen computers. There's also a very paranoia-inducing panopticon feeling to what we're seeing. As one person walks along, we see giant pictures of her (or her avatar?) leaping up the walls of buildings. And when she walks in to a store, she's shown pictures of herself in the clothes she might want to try on.[1]

Glass is an old technology. It is ubiquitous in the sense that it has been, especially since the nineteenth century, everywhere in our homes and cities (Armstrong 2008). It is a feature of twentieth-century transport and communications and is the material basis of scopic, screen and lens cultures. In the twenty-first century, glass is being redesigned as a more haptic, gestural and intuitive interface, neither more nor less visible than the screen or lens but more natural, closer in property first of all to plastic (which bends) and ultimately to skin (which stretches and breathes and has the semi-autonomous status of an organ). Annalee Nevitz marvels at how, in the Corning video, 'we see the touch screens working' and the devices networking as if they were autonomous, and this quest for, or at least appearance of, autonomy is what marks the point at which glass becomes an information technology, as it were, in itself. Glass as an information technology in itself retains its 'paranoia-inducing, panopticon feeling' of being seen always and everywhere, and it continues to present the scopic-haptic subject with images (projections and reflections) of ambiguous ontological status (Armstrong 2008). This same subject, moreover, continues to be gendered, coded feminine in contexts of narcissism and consumption. While 'she's shown pictures of herself in the clothes she might want to try on,' the glass itself becomes an agent of voyeurism and production; it shows pictures and it networks in order to produce information and render itself sensuously, desirably inconspicuous.

In *The Material of Invention*, design theorist and historian Enzio Manzini charts the coming together of matter and information technology that makes materials like glass appear autonomous and enables them to be

DOI: 10.1057/9781137374851.0004

branded – as they currently are – smart (1989). Glass contains all of the potentiality (and uncertainty) of *i*media. As an *i*material, it is becoming an *i*nvisible, *i*nformation *i*nfrastructure with its origins in the *i*nternet and its evolutionary trajectory oriented toward the organistic status of an *i*ntelligent skin. The teleology of glass is already indicated in experimental and environmental architecture and specifically in developments such as GreenPix, a zero-energy, solar powered transparent media wall 'for dynamic content display, including playback videos, interactive performances, and live and user-generated content' (Brownell 2010: 145). GreenPix's so-called 'intelligent skin' interacts with the interior and exterior environment of a building using embedded, bespoke software and 'transforming the building façade into a responsive environment' (145). The idea of intelligent, responsive, skin-like environments stems, as Manzini recognizes, from the introduction of artificial intelligence (AI) into design and architecture, and is perpetuated by the combination of artificial intelligence and ubiquitous computing in branded fields such as ambient intelligence (AmI) and augmented reality (AR). Corning's glass is smart in the AmI and AR sense, where AmI stresses the environmental, embedded, responsive, adaptive and predictive aspects of computing and AR remains consistent with a focus on actual rather than virtual environments that are simply overlaid or augmented with information (Furht 2011; Nakashima et al. 2010). Both fields are, I've argued, disingenuous in their claims to be human-centered and non-hubristic; their ostentatious stepping down and away from autonomous artificially intelligent life forms and virtual worlds a cover for transformative, even metamorphic, practices that turn scopic-haptic subjects into data objects for markets (Kember 2013). Corning's glass, like Google's or Microsoft's, is metamorphic in this way and by virtue of its increasing smartness (a performatively futuristic claim) it comes to effect changes not only on subject-objects and environments but also on itself.

The metamorphosis of glass into skin via plastic is facilitated by information technology and, Manzini suggests, by the suspension of the historical and cultural identity of glass in favor of its newly worked technoscientific properties, such as thinness and flexibility. It is, he says, no longer a question of what glass is but rather of what it does, even if its doings, its assembly of properties and performances remain concealed, hidden or obscured in the memory of what it was (29). It may even be the case that a nostalgia for the identity of glass as something cold, clear and breakable is heightened and made more necessary by its more plastic and elastic properties, its clarity a

DOI: 10.1057/9781137374851.0004

reassuring echo and response to the uncertainties and opacities it contains. Such an echo does not amount to an ontology of glass as an object or even as a process. For Manzini, there are only ever onto-epistemologies, materials comprised of narrative, symbolism and representation as much as matter. As these elements are worked together, glass remains clear – translucent or transparent – even as it bends without breaking and feels warm to the touch. Insofar as plastic is materially and symbolically about mutability, everything, including glass, is plastic now. The elasticity of glass, its aspirational organistic property, is facilitated by a combination of ubiquity and proximity to its subject. Adam Greenfield identifies two aspects of ubiquity to which we might add a third. For him, ubiquitous means everywhere and 'everyware' (2006). Everyware technologies are the products of AI and ubiquitous computing. They are embedded in objects, environments and materials that render them 'ever more pervasive and harder to perceive' (9). But everywhere and everyware technologies are set to be joined by ubiquitous everywear technologies including watches, glasses and lenses. These branded pre-products by Apple, Google, Samsung and so on remind us of the importance of Manzini's distinction between matter and material, the latter being worked, designed, instrumentalized and naturalized with reference to the former. They also remind us of the proximity of all, not only wearable technologies, or the degree to which human subjects and media and technological objects have always co-constituted one another (Kember and Zylinska 2012).

Materials scientist Mark Miodownik, echoing Marshall McLuhan argues that 'the material world is not just a display of our technology and culture, it is part of us. We invented it, we made it, and in turn it makes us who we are' (2013: 5). Like Manzini, he recognizes the limits and potential of disciplines in and across the two cultures that have so far struggled to account for human-technological relations. He rejects the scientism that characterizes, for example, object oriented philosophy and *i*media theory in its multi-scalar, dialectical mode – 'there is more to materials than science' – and takes the risk of combining cultural and technological accounts, science and story because 'materials and our relations with them are too diverse for a single approach' (8). Miodownik looks at glass as a material that bends light and transmits information. If, for him, it is – unlike wood, for example – still too cold and featureless for us to love in itself, it is its very smoothness and transparency that enables us to tolerate our proximity to it and its proximity to us, as our eyes and other boundary organs (179).

DOI: 10.1057/9781137374851.0004

Glass is a great leveler: democracy through transparency

Glass makes everything clear. In doing so, it obscures its own agency as a mediator and maker of worlds and things and obviates the need for further inquiry into them. Glass makes everything clear already, presenting and presencing the world and its subject-objects just as they are. As such, it obviates the need for further action as well as inquiry. Everything and everybody is already placed before us. By making everything clear to everybody equally, glass itself appears democratic. There is no further need for politics. Glass obscures the conflict and contradiction that inhere in the inequality between subject-objects for whom everything is equally clear. Properties of clarity, transparency, openness, accessibility, inclusiveness and democracy and above all, its ability to obscure conflict and inequality make glass if not loved then favored, sought after and invested in by institutions and organizations of power from governments to Google.

During the nineteenth century, glass, along with steel, was a cornerstone of the industrial revolution. Governments, commercial organizations and individual consumers invested in glass. (Blaszczyk 2000). It was the architectural and decorative material of The Great Exhibition at Crystal Palace, generating 'an environment of mass transparency never before experienced' as well as a whole new language and poetics of seeing and being seen (Armstrong 2008: 1). The Great Exhibition was open to everyone equally across the divisions of class, race and gender that the process of industrialization exacerbated and that the exhibition exhibited photographically, anthropologically, along with the collected and classified artifacts, products and other wonders of colonial exploration. Promising absolute knowledge for all through transparency, Crystal Palace was, for Isobel Armstrong, a 'descendent' of Jeremy Bentham's Panopticon – a model of surveillance and of knowledge as power (117). Glass itself might make everything clear to everybody equally, but its design and architecture, its cultural and technological working 'is never neutral' but rather imbricated in power and social division. If governments and Google are drawn to the ontology of glass – its essence as openness and democracy – the citizens of Crystal Palace and of twenty-first century glassworlds attend to the 'power relations of the window' and the prism, the hierarchical divisions of seer and seen, self and world, public and private space (117–9).

DOI: 10.1057/9781137374851.0004

If the nineteenth century was the 'era of public glass,' it is tempting to think of the twenty-first century in terms of the personalization of glass media (*i* as in *my*) and proliferation of private consumption. Here we might think of Jodi Dean's reference to the 'seeming dementia' of the individual walking along the street and appearing to talk either to themselves or to their phone, watch or glasses. However, divisions between public and private, though hierarchical, have never been stable. Henri Lefebvre commented on the over-decorated domestic Victorian window – 'swagged, draped, tasseled, frilled, ruched, layered with curtaining' – as a hysterical response, 'a fraught reaction to the abrupt termination of possession that it marked' (119). Thad Starner, one of the architects of the seemingly ill-fated Google Glasses – an everywear AR technology based on a light-bending prism – draws attention to the same inside/outside, private/public boundary and to fraught reactions that were potentially legislative and certainly decorative. Starner points out that new laws would be needed to secure the uptake of Google Glass and manage the privacy issues generated by what might have been (or might, in some form, still be) another descendent of Bentham's Panopticon[2]. Google Glass was intended to be a proximate everywear technology for everyone, but even before it failed to realize any popular appeal, it turned out to be for everyone who did not already wear glasses. The prism, combining visual information from the outside world with whatever was projected or layered onto it, directed a split beam onto the wearer's retina but was not initially adapted to long- or short-sightedness. Once it had been, Google Glass was to augur – along with the technologies that survive and will succeed it – another environment of mass transparency, another language and poetics of seeing and being seen that is based less on reflection, refraction and magnification and more on data visualization, facial identification, mobile location, gamification and life-logging. The undifferentiated and unequal citizens of Google et. al. will have equally clear access to all of the artifacts, products and other wonders of the augmented world.

*His*tories of manufacturing and magic

At this point, on the brink (treacherous for some, less so for others) of twenty-first century glassworlds, it is worth remembering Cinderella.

DOI: 10.1057/9781137374851.0004

Cinderella's slipper became an elastic glass one for the first time in nine-teenth-century versions of the story. She and it are now evoked in Google tales of domesticity and kitchen chores (Armstrong 2008; Schmidt and Cohen 2013). Cinderella comes in to *hi*stories of manufacturing and magic, of driverless cars and pumpkin carriages by Erik Schmidt and Jared Cohen and by Charles Perrault and George Cruikshank (2013; 1697; 1854). She comes in – bidden or unbidden – in order to differentiate the citizens of nineteenth-century and twenty-first century glassworlds alike. The *hi*stories she comes in to are concerned, specifically, with the transformations that connect manufacturing and magic and that have to do with glass.

Armstrong points out that before the last quarter of the nineteenth century, most mass produced glass was 'blown by human breath' and was an outcome of artisanal rather than mechanized processes (4). Glass was a combination of petrified breath and frozen liquid. Even the 'prefabricated panels of the Crystal Palace in 1851 were made up of 956,000 square feet of such breath-created glass' (4). It was only later in the century that the machine-made rolled sheet glass that is still being developed today[3] began to replace blown techniques (4). As Armstrong puts it:

> To look through glass in the mid-nineteenth-century was most likely to look through and by means of the breath of an unknown artisan. The congealed residues of somebody else's breath remained in the window, decanter, and wineglass, traces of the workman's body in the common bottle, annealed in the substance he worked.
>
> (Armstrong 2008: 4)

Traces of the body remained visible in the glass. They existed as tiny bumps and blisters, marks of 'bodily labour' and also of a brief life expectancy (5). Glassworkers of the nineteenth century did not live long and were often mythologized for their 'heroic acts of labour' (5). A consciousness of such feats as well as their material residue elevated glass to the status of an 'eternal substance,' a paradox of breath and matter and even of spirit and matter. The majority of glass artifacts 'arrived with a history of labour and transformation' embedded in the material, prior to its further transformation into the finalized product (5). The first transformation, that which is embedded in the material, occurs between spirit and matter and traces the movement between life and death. The trajectory of the laborer from life to death is in a sense mirrored and

DOI: 10.1057/9781137374851.0004

reversed by that of the matter he works. The matter that composes glass is predominantly dead stuff, sand and ashes:

> What can be meaner in appearance than sand and ashes… the furnace trans-forms this into the transparent crystal we call glass, than which nothing is more sparkling, more brilliant, more full of lustre. It throws about the rays of light as if it had life and motion.

> (Barbauld in Armstrong 2008: 5)

It is interesting, following this nineteenth-century poetic meditation on glass, that the language Anna Laetitia Barbauld uses is currently being used again. For Erik Brynjolfsson and Andrew McAfee, 'brilliant' technologies are the next smart technologies due in part to the lustre of their material (2014: 207). The continued transformation of glass, 'pure transparent matter derived from waste matter' confirms 'the magic of a transition from nature to culture' and even seems to 'reverse the process of mortality, moving from death to life' (Armstrong 2008: 6). This move-ment is completed, I'm suggesting, in the elasticization of glass as skin and in the return to nature from culture.

The second transformation, from brilliant material to commercial object, is made to seem 'almost invisible' due to the transparency of glass. Here, manufacturing is made magic by what the nineteenth-century novelist Anthony Trollope refers to as the 'double lustre' of glass, 'giving aura to ordinary objects and multiplying the allure of luxury goods' (in Armstrong 2008: 121). A twenty-first century environment of mass transparency – one that is very much desired and in the process of being designed by Corning, Google, Microsoft et. al. – depends on this double lustre and on the utterly clear, no questions asked exchange value of alluring goods such as smart phones that are constantly upgraded and never permitted to become ordinary en route to their ubiquity. But even the faded lustre of a product that has been superseded, downgraded and turned back from its magical to its merely manufactured status and then replaced by a more sparkling and brilliant version cannot account for the ultimate transformation that inheres in *his*torical glassworlds. The ultimate transformation has to do with things that are turned into other types of things: pumpkins into carriages for example, or lizards into coachmen and scullery maids into princesses. With all of this bound-ary crossing between animals, vegetables and humans and between animate and inanimate things, the question that arises, as Armstrong suggests, is 'whether transformation is a transgression.' The figure of the

DOI: 10.1057/9781137374851.0004

metamorphosed body 'co-opts Cinderella into the grotesque imaginary where the elastic glass slippers, as matter, breath and artifact, also belong' (2008: 210).

Armstrong reads the elasticity of the nineteenth century glass slipper in terms of the transition from the magic of glass to its manufacture. The elastic slipper is no longer magical, no longer pure because, by the nineteenth century, storytellers simply knew too much about how glass is worked. I read the elasticity of twenty-first century glass – its aspirational organistic property – backwards from there, from the merely manufactured to the magical and from 'modernity's way of transforming things into something else' (materials into products) back to their mythical status. For me, the prospect of an elastic glass slipper is grotesque precisely in that it describes a metamorphosed body, the transformation of glass into skin. In as far as glass effects transformations *on* subjects, objects and environments, it does so, increasingly, by a process of becoming-with them. Glass is certainly the principal agent of Cinderella's transformation. A 'feminine vessel,'[4] the glass slipper does its magical-manufacturing work in a patriarchal context concerned with conspicuous feminized consumption and 'sumptuary excess' of the sort that is embodied in the idle, corpulent, narcissistic and markedly over-dressed forms of the ugly sisters. But the glass that works on Cinderella, turning her into a princess fit for a prince has, as Armstrong suggests, always shown signs of working on and transforming itself, of crossing boundaries and becoming something else like breath – or breathing. As it becomes more ubiquitous and proximate, glass starts to work not on but *as* Cinderella. It does so in a contemporary patriarchal context concerned with conspicuous feminized productivity and potentiality and with labor that is efficient, flexible, service-oriented and nurturing if no longer heroic.

Cinderella glass

Corning's advanced glass is Cinderella glass. While it dreams of becoming so much more than it currently is – namely, 'a critical component of today's interactive, device-oriented, connected world' – it simply works. In fact, 'glass does more and more of the "work" of the display in the device,' enabling complex electronic circuits with greater than ever functionality.[5] Cinderella comes in unbidden, un-cited in Corning President

DOI: 10.1057/9781137374851.0004

James Clappin's story of Willow Glass™ that 'will help enable thin, light and cost-efficient applications including today's slim displays and smart surfaces of the future.' Gone are the days of narcissism and sumptuary excess. Cinderella now works for a different kind of sister, altogether more productive and streamlined. The moralism directed at body shape and size is barely concealed. It is the 'thinness, strength and flexibility' of Cinderella's own body that enables her, in the form of Clappin's Willow Glass™ 'to be "wrapped" around a device or structure, potentially revolutionizing the shape and form of next-generation consumer electronic technologies.'[6] Corning's advanced glass is 'ultra-slim,' 'flexible,' 'conformable (curved)' and skin-like, wrapping around – covering and protecting – the new generation of intelligent artifacts it gives life[7] to. Willow Glass™ is re/productive, high-performing and organistic, with 'natural hermetic properties that make it a seal for organic LED (OLED) displays and other moisture-and-oxygen-sensitive technologies.'[8] Clappin differentiates Willow Glass (tough) from Gorilla Glass (tougher) and associates its 'thin as paper' svelteness and flexibility with the future of transparent, OLED devices such as phones and electronic newspapers – that allow light to pass in both directions – as well as wearables.

Glass remains clear despite its more plastic and elastic properties. As well as working for its attractive, streamlined and smart sisters, it works to unmediate a future fantasy world it is already becoming-with; an already metamorphic, augmented, ambiently intelligent world of neutralized, undifferentiated data subjects and seemingly autonomous networking things-in-themselves. For its progenitor, Sergey Brin, Google Glass was intended to be the very model of unmediation, an attempt (failed for now) to erase *i*media as the in/visible creators of *i*worlds. For Brin, Project Glass was all about direct, immediate access to *i*worlds and the wonders they contain. The glasses were designed to erase their contribution to these worlds, to mask their part in making them up by being ubiquitous and proximate. They were intended to be and may, in some guise yet be everywhere/ware/wear as our eyes and ears, directing a split beam of information input and output, projection and reception onto our retinas and bypassing other organs altogether by attaching sound sensors to our heads. This, Brin declares, is 'freaky but you get used to it.' It seems, in the end, that we didn't but Brin's point is that the workings of our senses and prostheses, their constructions and representations of the world are simply too intrusive or rather, defensive. For him, they should be removed, undone by unmediating technologies that

DOI: 10.1057/9781137374851.0004

restore our courage to be in and to face up to the world as it is. The moral imperative is as clear as the glasses themselves and again, Cinderella is evoked and embodied in the form of feminizing, 'emasculating', glass-wrapped smart phones, the fiddly, attention-seeking pathetic prop that smart glasses supersede, allowing and obliging proper men and princes to be prop-less, to face up rather than look down, to re-enter the world anew, wearing little more than their intelligent skin.[9]

Historians and theorists of glass as a material and as an *i*material remind us that glass has always been working towards the endpoint of unmediation and doing so in a way that is opaquely gendered, that associates immediacy with masculinity and by extension, 'hypermediacy' – the all-too visible signs of communicative procreation – with femininity (Bolter and Grusin 2002). But glass is actually an 'antithetical' in/visible material, one that 'holds contrary states within itself' (Armstrong 2008: 11). Glass has its own double logic[10] as a form of unmediation, making itself seen and felt precisely in the attempt at self-erasure. As Armstrong points out, 'we would not call it transparent but for the presence of physical matter, however invisible – its visible invisibility is what is important about transparency' (11). Transparency, or rather, the claim to transparency reveals its own opacities, as Claire Birchall and others have pointed out with particular reference to governments and Google (Birchall 2011; Hall 2012; Phillips 2011). What is revealed and what is hidden are the constitutive outsides of power and of its chosen, its most favored material. The 'natureculture' or 'onto-epistemology' of this material is ultimately its clarity in ambiguity (Haraway 1991; Barad 2007). The ambiguity of glass goes all the way from its liquid-solid state to its ability and tendency (culturally, historically, materially) to at once give and withhold information, including, perhaps increasingly, about its own *i*material agency. In contexts of design and architecture and in stories of manufacturing and magic, glass 'ceases to be that which is perfectly clear and becomes, instead, that which is clearly ambiguous' (in Brownell 2012: 106). This constitutive clarity in ambiguity is both a condition and an affordance of glass. If it cannot fail to be recognized, it can be exposed, exploited and experimented with to a greater or lesser extent (Brownell 2012). There is, at most, an ironic allusion to the gendered opacities and ambiguities of advanced glass in Clappin's choice of nomenclature – the apparently straight-faced exposition of Willow Glass™ knowingly juxtaposed though never quite undermined by reference to the 'sort of abuse that the more brolic Gorilla Glass is

DOI: 10.1057/9781137374851.0004

built to withstand' – but such irony only reinforces their existence. Armstrong and Blaszczyk provide the back story to Willow Glass in their sense of the gendering of cut glass bowls, window reflections and, of course, glass slippers in the nineteenth century (2008; 2000). It is no coincidence that they both draw on Cinderella as a *his*tory of clarity in ambiguity that is forever being retold and that, like glass itself, is therefore open to experimentation and transformation.

Remember Cinderella

Remembering Cinderella means seeing and touching the clarity in ambiguity of glass, recognizing, in the double lustre it confers on things, an always already grotesque ambition. This is not, of course, the ambition of glass itself but that of its manufacturers and magicians – storytellers all. Those who recognize the ambiguity of glass and even those who play with it do not take sides and cannot be clearly identified as either technocapitalists or academics, architects or writers. Glass is an imaterial for everyone, albeit unequally so that Clappin, for example, or Cruikshank are able, through the legacy of patriarchal capital, to proffer versions and visions to contest. Just as *Cinderella* continues to be a contested as well as multi-scalar tale, Cinderella glass – as a futuristic fantasy figure of feminine and feminized labor magically transformed into something or somebody else – remains, if not up for grabs then subject to diffraction. Diffraction is not the continued, iterated reflection, refraction and magnification of women in and as glass but rather a mode and method, as well as a phenomenon of interference that interrupts and redirects it. Retold stories count as diffraction, whether they are genealogies or fictions, her stories of transparency and gendered opacity in decorative, domestic, functional and working glass or feminist science fictions. Blaszczyk, like Armstrong, is a genealogist of glass. Her story is about the *his*toried alignment of Corning with the science of home economics and feminized consumption (2000). The material semiosis of streamlined efficiency and flexible, bendable, spotlessly clean and clear domestic productivity can be traced back to the nineteenth century and forward again, at least as far as the 1950s, when Corning's Pyrex glass, marketed to women, worked toward making them better, more professional home workers. Marina Warner, like Angela Carter, deals in what Haraway terms feminist sf,

DOI: 10.1057/9781137374851.0004

an antagonistic if not oppositional mode of remaking up the made-up world that combines gendered facts and fictions (Haraway 2011). Carter's *Cinderella* is a translation of Charles Perrault's and, at the same time, a reworking of it, one that gives magical powers to godmothers rather than glass (2008).[11] Warner's *Cinderella* is a fantasy figure of patriarchal capital that enrolls women story tellers, including Carter, in misogynistic tales of 'hapless heroines,' good fairy godmothers and wicked stepmothers and in performing paranoia, the splitting of good and bad mothers, rather than engaging the historical conditions that set women against women in the home (1995: 207). My own Cinderella is an sf figure from and beyond the fairytale. She is a figure of manufacturing and magic joined by glass and morphed, grotesquely and transgressively from plastic to skin to somebody real and imaginary. Remembering Cinderella enables me to tell a retold story of *A Day in the Life of Janet Smart*, my version of Corning's vision of *A Day Made of Glass*, Microsoft's original *Future Home* and *Productivity Future Vision* and even Google's *Project Glass: One Day* (as told by Sergey Brin) all of which feature a generic woman in/as her Cinderella glass.

Notes

1   Annalee Nevitz 'A beautiful but creepy vision of the "smart glass" future' http://io9.com/5781931/a-beautiful-but-creepy-vision-of-the-smart-glass-future

2   https://www.singularityweblog.com/thad-starner-google-glass-augmented-reality/

3   Corning are developing a roll-to-roll rather than sheet-to-sheet production process https://www.corning.com/in/en/products/display-glass/products/corning-willow-glass.html

4   Blaszczyk examines the gendering of glass objects such as the cut glass bowl and looks at the promotion of pressed glass for women versus cut glass for men in nineteenth century America (2000).

5   http://www.aml-tech.com/Glass-and-Ceramics-A/Benefits.html

6   https://www.corning.com/in/en/products/display-glass/products/corning-willow-glass.html

7   That is, life-form but also, in historical terms, breath.

8   https://www.corning.com/in/en/products/display-glass/products/corning-willow-glass.html

9   https://www.ted.com/talks/sergey_brin_why_google_glass?language=en

DOI: 10.1057/9781137374851.0004

10   For Bolter and Grusin, immediacy and hypermediacy constitute the double logic of remediation (2002).

11   Crucially, Carter also reworks Perrault's moral reading of the tale that for her states that whatever your talents, 'they will never help you to get on in the world unless you have either a godfather or a godmother to put them to work for you' (2008: 37).

DOI: 10.1057/9781137374851.0004

3

Ubiquitous Women: Everywhere, Everyware and Everywear

Abstract: *Ubiquity, like complexity, is too easily and perhaps too often regarded as a thing-in-itself – an objectively occurring phenomenon of iworlds rather than a mechanism for making them or even remaking them in the image of the past. This chapter examines ubiquity as a mechanism for remaking the past in the future and, specifically, for containing potentially unlimited possibilities for women in domestic and urban environments in which they are everywhere (hypervisible, luminous), everyware[i] (bound up in the lustrous luminosity of smart glass technology) and everywear (in wearables whose regulatory role is oriented to the reproduction of women as sexual objects and hyperproductive subjects). Through the gendered configuration of ubiquity, technology and time, I highlight not just the remediation of images and ideals from the 1950s in futuristic industry visions of the early twenty-first century but the strategic value of indecision, or undecidability in feminist time-telling. If it is the case that ubiquitous women emerge both in time – historical time, clock time, structured time, productive time, optimized time – and as time – where time is understood as creative evolution or as life itself – then it is necessary, I suggest, for feminist time-telling to incorporate conflicting concepts of women as potential and as potentia. I argue that potential and potentia are co-constituted and co-constitutive of ubiquitous women. It is not possible to choose between them, or to carry on choosing between them on the basis of a feminist ethics or politics.*

Keywords: Feminist ethics; feminist politics; technology; life; time; ubiquity

Kember, Sarah. *iMedia: The Gendering of Objects, Environments and Smart Materials.* Basingstoke: Palgrave Macmillan, 2016. DOI: 10.1057/9781137374851.0005.

DOI: 10.1057/9781137374851.0005

A day made of glass

Jennifer is asleep in bed with Jeremy.[2] His arm is draped across her. He rolls on to his back, eyes closed, grinning. Behind them there are glass doors – 'PHOTOVOLTAIC GLASS. High Efficiency, Optically Versatile, Durable' – and at their feet is a large glass screen – 'LCD TELEVISION GLASS. Large Format, Ultra Thin, Frameless Design' – that wakes them at 7 a.m. with birdsong. Jeremy sits up, stretches and goes to touch the screen. With a flick, he changes the 'LIVE SKYCAM' view to the news. In the bathroom, Jennifer turns on the tap. It is large – an oddly, incongruously phallic close-up. Her mirror opens five separate windows: live news, weather, temperature, today's schedule (she has an executive brief, a project meeting with production and a brainstorming session with the team all before 10 a.m.) and automated applications (including shower settings and coffeemaker). Jennifer regards her schedule with a concerned expression. She selects 'project meeting with production' and another window reads: 'Hey Jennifer, 9.30 is now 8.30. Can you make it?' Still in her pajamas, she goes to respond by touching the 'ARCHITECTURAL DISPLAY GLASS. Pristine Surface, Electronics Enabling, Touch Sensitive' and dragging up her virtual keyboard. Here, she types 'ok, yes' and, with a slight flourish, sends. Jeremy is now dressed all in blue. He shows up nicely in the gleaming white kitchen. He chops peppers and scrambles eggs while watching the morning report on the worktop surface. He adjusts the hob temperature with his index finger and starts cooking breakfast on the 'ARCHITECTURAL SURFACE GLASS. Tough, Thermally Durable, Display Enabling.'

Jennifer and Jeremy's two girls appear, and one of them dumps a pair of pink trainers on the glass worktop. She enlarges a digital image of herself and her sister 'pinned' to the fridge door. This is 'APPLIANCE VENEER GLASS. Seamless Design, Electronics Enabling, Scratch and Smudge Resistant.' Tough for him, scratch and smudge resistant for them. There is Corning glass for all the family. The family is mixed race. Jennifer is white. Jeremy is Asian American. The younger sister draws cat whiskers on her older sister's photograph. Other photographs turn into videos. In one, the girls drape their arms around each other. In another, they make faces at the camera. There is a transparent glass phone on the worktop behind Jeremy. It rings and he picks it up. It's his mother calling (at 7:17 a.m.). He accepts the call by touching 'HANDHELD DISPLAY GLASS. Thin and Lightweight, Damage Resistant, Touch Sensitive.' His

DOI: 10.1057/9781137374851.0005

Mom's image enlarges and jumps out when he places the phone back on the worktop. He passes her to the girls for a chat.

Jennifer comes in dressed for work, grabs an apple and heads for the car. Her car greets her with a display unit that reads: 'Hello, Jennifer.' She sets the navigation to Main Office and pulls up directions (maybe she's new) on the 'AUTOMOTIVE DISPLAY GLASS. Streamlined Design, Advanced Functionality.' She responds to a reminder about her production meeting, courtesy of her 'AUTOMOTIVE DESIGN GLASS. Photosensitive, Durable.' Then she heads downtown, glancing up at a roadside display on 'LARGE-FORMAT DISPLAY GLASS. Weather Resistant, Electro-optics Enabling, Optically Adaptable.' The display transfers to her car, suggesting alternative routes to work that avoid an accident roadblock. Downtown, she passes a glass bus stop which displays directions to 3rd Avenue in English and Chinese. The stop is made from 'ALL-WEATHER SURFACE GLASS. Damage Resistant, Display Enabling, Touch Sensitive.' At the stop, an African-American man is talking to three people on his transparent phone. We join one of them in the office, where he works as a fashion designer. The bus stop man now appears, along with some colleagues, on the office-based 'WALL FORMAT DISPLAY GLASS. Large Scale, Seamless Design, Touch Sensitive.' Alongside the live screen captures from London (4:43 p.m.), Seattle (8:43 a.m.) and New York (11:43 a.m.) are some drawings of female models combined with digital fabric swatches and photo shoots, circled in red. The New Season Media Campaign is reviewed by placing a transparent glass phone on a display surface made of glass. The projected image includes a live model. She is manipulated, made to swivel and change clothes by the touch of the designer's finger. This is 'WORK SURFACE DISPLAY GLASS. Durable, Versatile, Application Enabling.' The designer pushes an enlarged video of the model onto the wall display of 'ELECTRONICS-READY GLASS. Pristine Surface, Exceptional Optical Clarity, Interactive.' The camera zooms in towards her eye and then, seemingly, through this to an electronic circuit board, her newly visible cyborg interior.

Meanwhile, Jennifer pulls into town (not the office) and a huge Times Square display of the New Season's fashion bears down on her. The display – on 'LARGE PANE DISPLAY GLASS. High Visual Impact, Multifunctional, Interactive' – plays inside as well as outside a glass building and, in case she somehow misses it, also on her phone. She receives a message to drop by and check out the new collection, and

DOI: 10.1057/9781137374851.0005

she does so, walking in to the shop as a virtual model approaches her, life-sized and in a purple dress. Jennifer is welcomed back by name and invited to browse the store. She chooses from the following categories: 'Dresses, Cardigans, Casuals, Accessories.' Selecting casuals, she is presented with five pint-sized models dressed in cardigans and slacks. She seems pleased, her smiling face reflected in the glass as the miniature models turn in front of her. She picks the one wearing an Appliqué Ruffle Blouse and spins her around by swiping her fingers across a bar on the screen. Bus stop man arrives in person at the office and unfolds a roll of 'FLEXIBLE DISPLAY GLASS. Ultra Thin, Rollable, Electronics Enabling.' He explores some architectural plans while Jennifer checks out her very own Appliqué Ruffle Blouse in the mirror.

We fast forward to the evening, never having seen Jennifer at work. She is, in fact, at home, carrying a bowl of popcorn and wearing her new top. Jeremy takes some popcorn and kisses her on the cheek. She sits down on the sofa with the kids who are watching something cosmological on the '3D TV DISPLAY GLASS. Vivid and Immersive, Thin and Lightweight, Frameless Design.' As they watch the planets orbit around them, Jeremy lies in bed reading H.G. Wells' *The Time Machine* on his 'PORTABLE DISPLAY GLASS. Ultra Thin, Flexible, Electronics Enabling.' (Corning, A Day Made of Glass, 2011)[3]

The second – same day, expanded – part of Corning's story of a day in the life of glass is about 'Enabling A Future of Communication, Collaboration, and Connectivity.' The focus, at the outset, is Jennifer's daughter, who is woken at 7:05 a.m. to floating images of her friends that emanate from her tablet or 'ELECTROCHROMIC GLASS. Privacy Enabling, Colour Selectable, UV/Thermal Insulated, Durable.' Jennifer's daughter is given five windows projected onto a glass wardrobe: profiles of her friends; weather; school schedule (a reminder about her field trip); a set of recommendations (including the application of sunscreen) and a choice of clothing. The 'ARCHITECTURAL DISPLAY GLASS. Durable, Seamless, Electronics Enabling, Touch Sensitive' presents her with an 'Outfit Explorer' to help make sure that everything matches. Jeremy drives the girls to school and they hijack his 'Dashboard Themes' with pink hearts and start singing. Jeremy laughs it off and puts something more practical back on his 'AUTOMOTIVE DISPLAY GLASS. Custom Contoured, Glare and Smudge Resistant, Touch Capable.' The school roof is covered with 'PHOTOVOLTAIC GLASS. High Efficiency, Durable, Lightweight, Hermetically Sealed' and the school walls display

'Park Elementary Energy Use.' The next generation of Corning glass is not just smart but more environmentally friendly. The children at Park Elementary each have a transparent tablet, and today's lesson looks like an information orrery with the weekly schedule, attendance and homework, announcement and science lesson 'planets' orbiting around the home room star. The teacher selects a subject – light – from the 'WALL-FORMAT DISPLAY GLASS. Durable, Seamless Wall-Size Coverage, Touch Sensitive' and so begins another orrery of optics, color and energy all whizzing around the symbol of a light bulb. The teacher gives everything a push, and the children in this information playground appear suitably impressed. They engage with a large color spectrum projected onto 'WORK SURFACE DISPLAY GLASS. Durable, Multi-Touch Enabling, Pressure Differentiating' and get to see lots of images of things that are blue.

Meanwhile, Jeremy, who is on a hospital ward wearing a white coat, talks to an African-American woman carrying 'SPECIALITY TABLET GLASS. Anti-Microbial, Scratch and Chemical Resistant, Thin and Lightweight.' He flicks information across to her tablet from his and enters a pristine surgery where he connects to China and talks to his virtual male colleague, Dr. Peng Zheng, by means of 'WALL FORMAT DISPLAY GLASS. Durable, Seamless Wall-Size Coverage, Touch Sensitive.' Jeremy can see Dr. Zheng's patient lying on a surgical table in China while he studies an MRI of his brain. Jeremy is a brain surgeon. Jennifer went shopping for clothes. Like Tom Cruise in *Minority Report*, Jeremy commands a wall full of visual data displayed on 'GLASS OPTICAL FIBER. Highest Bandwidth, Wireless Integration, Bend Insensitive.' At the same time, a female nurse in China uses 'ANTI-MICROBIAL GLASS. Multipurpose, Anti-Bacterial, Anti-Viral, Anti-Fungal.' This glass is clean! Like Superman, Jeremy sees through the virtual patient's skin and skull, the '3D-OPTIMIZED GLASS COMPONENTS. Superior Transmission, Distortion Free, Thin,' allowing him to study the brain as if it were on the outside. He can even remove slices of the brain with his finger.

Jeremy and Jennifer's daughter is on a field trip to Redwood State Park, bordered, of course, with 'ALL-WEATHER SURFACE GLASS. Display Enabling, Damage Resistant, Touch Sensitive.' The park ranger travels back sixty-five million years so that, just like in *Jurassic Park*, a dinosaur can appear and make them scream. Glass education is very like entertainment. The children run around identifying things by holding

DOI: 10.1057/9781137374851.0005

up 'TRANSPARENT DISPLAY GLASS. Augmented-Reality Enabling, Durable, Lightweight.' They take these things home to show Jennifer, projecting them from the sofa onto the '3D TV DISPLAY GLASS. Vivid and Immersive, Frameless Design, Next-Gen-High-Definition.' The three of them are eating popcorn again. (Corning, A Day Made of Glass 2: Same Day. Expanded Corning Vision, 2012)[4]

In the story behind its stories of glass, Corning takes us behind the scene of Amy's room. Amy is Jennifer and Jeremy's daughter. The narrator and guide of 'A Day Made of Glass 2: Unpacked,' mentions Amy's switchable (transparent to opaque) smart glass bedroom window, since this is one of the few glass technologies on the market.[5] For Corning's multiple and multiplying categories of ultra-thin, durable, pristine, flexible, tough, electronics-enabling, interactive, 3D, etc., display glass to become a reality, it must keep working with its partners to develop the relevant operating systems and apps. We learn nothing about these, but do find out that Amy's sister is called Sarah and that 'in tomorrow's world, configurable touch dashboards will be the norm.' Display glass will in fact be everywhere. When? It is available now, 'but not at this scale and not at an affordable price.' Corning states that there needs to be more innovation in manufacturing before it can fulfill its visions. Accessing sufficient bandwidth is an issue, and in order to hide all of the electronics, those big sterile and spectacular walls are more likely to be reduced to the size of relatively small screens. There are, in short, obstacles to overcome. Somewhat disenchanting for the viewer, these are clearly a frustration for Corning, dependent as it is on the manufacturing industry and on massive improvements in global communications infrastructure. (Corning, A Day Made of Glass 2: Unpacked. The Story Behind Corning's Vision, 2012)[6]

En route to the third version of its vision, Corning offers some high school students' reactions to the first two versions. Here, they opt for an obstacle and explanation free drama of before and after viewing, and to my mind, the more interesting student comments are made before they watch the videos. These are oriented to the properties of glass, its clarity and smoothness, its durability, transformability and apparent antithesis to technology.[7] What Corning subsequently reveals is its own effectiveness as a storyteller and how effective stories themselves are at in-forming their audience, writing them in to the futures that are told, in this case about all aspects of daily life from domestic to educational and professional and back again. Every child sees itself somewhere in this cycle and engages in the possibilities of how things could, one day, be.

DOI: 10.1057/9781137374851.0005

A future productivity vision

Corning supplies glass to the electronics industry, among others. Microsoft's vision incorporates glass along with other materials and is focused on the issue of productivity and its alignment with future glass worlds. Its story begins with a smartly dressed businesswoman arriving at an international airport. As she waits for her driver, her glasses translate the tannoy announcement at the interactive pick up zone. Inside the car, her translucent phone informs her that it is 4 a.m. in Johannesburg and displays a handwritten message: 'Miss you, Mom.' She draws a heart and, via her device, pins it to the kitchen wall at home. She looks out of the augmented reality enabled car window which counts down the time to her arrival at the hotel and highlights a building: 'APPOINTMENT LOCATION. YOUR MEETING TOMORROW IS HERE.' We learn her name via the Travel Hub and hotel check-in on her phone: 'WELCOME BACK AYLA KOL!' At the hotel, an African-American male bellhop assistant (Ayla is white) looks at the guest arrival queue on his transparent glass card, swipes the back of it with his finger and brings up a photograph of Ayla and her time of arrival. The back of the card gives comprehensive information about hotel guests: language and destination (English, Australia), service level (Platinum) and favorite amenities (yoga, pool, extra pillows). After a nine-hour flight and still in the back of a very dark car that enhances the lustre of her glass phone, Ayla is contacted by a colleague who wants her to review a proposal as soon as possible. She agrees to do so first thing tomorrow (or for her, later today) and as she arrives at her hotel, we cut to the colleague, Qin, at an underground station with a remarkably good wifi connection. Qin uses his translucent phone to review data visualizations about a sustainable building project. He approves an order and donates to a benefit concert by holding his phone up to a live visual display on the underground station wall. The performer appears to nod in recognition of his generosity. Qin is able to view all of the pledges, by station, in 3D. Cutting back to Ayla in her hotel room at 7:30 a.m., we see a number of the glass types classified by Corning: surface displays, wall displays and so on. Microsoft has added a holographic function so that the cloud that connects Ayla to her proposal actually does float in the air. The 'Greenwall Proposal' appears on a thin, flexiglass tablet next to Ayla's busy morning agenda. She cuts, pastes and connects using a combination of touch, gesture and voice-activated software. Ayla is only ever a virtual presence at the

DOI: 10.1057/9781137374851.0005

project headquarters where Qin (male) joins the project coordinator (also male) to review all manner of visualized data and ensure that the proposal is geared not only to sustainability but to maximum efficiency. Like Jennifer, Ayla never actually gets to work, despite having a professional profile and a hectic schedule. Ayla's return home is signaled by the presence there of her daughter, working on her maths at the kitchen table next to the wall with the 'Miss you, Mom' note and the heart. Ayla's husband or partner comes in to help his daughter, Shannon, make something for the bake sale. Ayla, still in her hotel room, joins in. She and Shannon choose a recipe from South Africa, which appears on the kitchen worktop back home. (Microsoft, Productivity Future Vision, 2011)[8]

I am reminded of an earlier, now obsolete version of Microsoft's Future Home promotional video. This featured a woman called Janet and a smart, speech-enabled kitchen that asked her if she wanted help with her baking. Janet said that she did and placed a bag of flour on a hotspot on the kitchen worktop that then displayed a recipe for focaccia bread. Then the cameras in her smart ceiling read the tag on her medicine bottle and the kitchen said (verbally and in text): Janet, don't forget to take your medication. A male character subsequently replaced Janet. I don't think he is medicated. As I've suggested elsewhere, Janet is all-too reminiscent of the housewife of the 1950s, the very housewife who features in Monsanto's vision of a future home made of plastic[9] and in Betty Friedan's account of the feminine mystique (Kember and Zylinska 2012; Friedan 2010). The housewife, for Friedan, is a figure of containment as well as regression to an earlier era before women went to work for the war. The housewife figure signifies the frustrations and ills of domestic enclosure, the waste and the cost of a society shutting down the possibilities for women at the very moment, and to the very extent that they were opening up. The presumption now is that these possibilities are fully open and that the work of feminism is done (Bates 2014; McRobbie 2010). I'm suggesting that this claim to openness as unlimited possibility materializes in the translucency and transparency of glass and in the clear – glass reflected, refracted and magnified – ubiquity of women everywhere (at home and abroad although somehow still at home) and everyware, their lives enhanced and augmented by information technologies seamlessly and in-visibly integrated in glass. The category of women is also subject to the claim of openness. Industry-driven visions of the future, in the second decade of the twenty-first

DOI: 10.1057/9781137374851.0005

century, signal their openness through the inclusion of generic differences – Asian women (and men), African-American women (and men), sometimes even slightly older women (and men) – while the central character remains resolutely young, white and middle-class, her role ultimately as naturalized as the material that surrounds her, working for her, working in place of her so that she can return 'home' to her traditional roles in her markedly heteronormative family.

Like Corning's, Microsoft's reactionary vision of a future that looks more like the past is consistent with the housewife-turned-working-mother as a figure of containment through openness – or rather, through the enclosures that are the constitutive outsides of openness (Kember 2014). Glass technology opens out onto a world of possibilities – social, cultural and communicative, economic, ecological, private, professional – that, like glass itself, cease being fluid and become instead increasingly delimited and demarcated; novel variations and categorizations, perhaps even new ways of marketing the same old thing. Glass is an old technology and the housewife and working mother is an old regulatory technique. Angela McRobbie picks up on Gilles Deleuze's reference to luminosity, a softer, more insidious form of panopticism, in order to account for the heightened visibility and scrutiny of young women with particular reference to their future productivity (2010). Regimes of self-management, care and improvement – both Ayla and Jennifer have prominent health check apps – ready them for a presumably open, accessible and egalitarian home (Jeremy cooks) and workplace which, in the case of Ayla and Jennifer (Janet never leaves the house) magically returns them before they even get there, turns them back into housewives, consumers and mothers, retaining only their grotesque, metamorphosed substitutes; their digital doppelgangers, virtual skins and luminous-lustrous glass presence. This lustrous luminosity, simultaneously adding value to products and soft-focus, soft-porn scrutiny to women is ubiquitous in the sense of being everywhere, everyware and also everywear.

Women, wearables and work

Part 1: Bras and pants, glasses and watches

Jennifer and Ayla's professional profiles, daunting to-do lists and numerous contacts are prominently juxtaposed with their health-check apps.

DOI: 10.1057/9781137374851.0005

Health-checkers, fitness trackers, smartbands and so on are associated with the phenomenon of lifelogging or self-quantification which revolves around the sanctioned, increasingly obligatory obsession with measuring personal data. The close monitoring as well as sharing of such data 'is part of a growing trend of "the quantified self" movement with the motto: "self-knowledge through numbers"' (Adams 2014: 13). Tim Adams suggests that the quantified self movement is rooted in the belief that 'the examined life' can be associated with 'data points' such as 'blood pressure, heart rate, food consumption, sleep,' quality and quantity of exercise as well as the extent of our 'real and social media interactions.' Once we add these up, he continues, we know who we are. In due course, this quest for self-knowledge through numbers, or more precisely through algorithms and algorithmic self-regulation will extend, via gamification techniques, from contexts of leisure to those of work. Such techniques, made manifest through workplace gameplay scenarios, are geared towards measurably enhanced productivity, performance, transparency and efficiency as the core values of neoliberalism. One company, Knack, 'uses immersive online video games to collect data to quantify how creative, cautious, flexible and curious' employees and potential employees might be (Adams 2014: 12). They offer predictive analytics about universalized, neutralized data subjects, their privacy intrusions characteristically justified by a humanistic, moralistic and ultimately economistic refrain about the optimization of human potential. However, the data subjects that constitute the apparently universal category of the quantified self are, I'm suggesting, increasingly, visibly re-differentiated in the forms – predominantly young, white, heterosexual and middle-class – of ubiquitous women.

The rising prominence, visibility and ubiquity of women in environments of imedia that incorporate and exceed popular culture – to include, for example, promotional videos by glass developers and the electronics industry that go viral on YouTube – coincides with and, as McRobbie suggests, is predicated on the 'disarticulation' of feminism from other social movements and, ultimately on its undoing (2010: 24). McRobbie's notion of disarticulation 'gives depth to the notion of anti-feminist backlash' and captures the reversal of a process, described by Laclau and Mouffe, whereby progressive social movements forged connections and alliances (25). Based on a 'post-structuralist and psychoanalytic understanding of identity as never transparent, never fulfilled or authentic, and never wholly corresponding with the structural categories of class, race

DOI: 10.1057/9781137374851.0005

or sex, they argued that lived social identities were always formed from a range of unstable and historically contingent elements' (25). It is this reading of identity, coincidentally, that is forgotten, or simply abandoned in the turn from a supposedly humanistic, subject-centered critique to the ontology of objects and processes. This turn is, as I've suggested in previous chapters, compatible, if not complicit with the disarticulation of intersectional or trans politics, a disarticulation that is 'the objective of a new kind of regime of gender power' based on 'forcing apart and dispersing subordinate social groups who might have possibly found some common cause' and rendering the possibility of a new feminist political imaginary 'inconceivable' (26). While recognizing that alliances were never unproblematic, McRobbie maintains that the 'cutting off' of feminism isolates, undermines and disarms it while precluding its future development. In place of feminist futures, she argues, 'there is an over-supply of post-feminist substitutes from within the new hyper-visible feminine consumer culture' (26). Even to the extent that feminism is accepted, reintroduced into mainstream, educational and political culture, there is always, she suggests, a price to pay, a punitive response that effectively contains and curtails the possibilities of feminist ethics and politics through, for example, the individualized injunction to self-perfection.[10] For McRobbie, writing in *The Aftermath of Feminism*, and for Laura Bates, writing about everyday sexism, the substitution or effective negation of feminism is marked by a veneer of equality with respect to educational, employment and consumer opportunities and by knowing, ironic, jokey forms of sexism often passed off as harmless banter (2010; 2014):

> This word, 'banter,' has become central to a culture that encourages young men to revel in the objectification, sexual pursuit and ridicule of their female peers – it is a cloak of humour and irony that is used to excuse mainstream sexism and the normalization and belittling of rape and intimate-partner violence. And it is incredibly effective, because – as we know – pretending that something is 'just a joke' is a powerful silencing tool, making those who stand up to it seem staid and isolated. But, as one young woman I interview explains: 'I don't find it funny.'
>
> (Bates 2014: 140–1)

The post-feminist, disarticulated, generically differentiated visions of the future presented by Corning and Microsoft offer a veneer of equality by following girls to school and women to work while foregrounding their consumer opportunities in the new economy of shiny glass

DOI: 10.1057/9781137374851.0005

communication technologies as in the old economy of blouses, kitchens and cakes. Their ubiquitous glass presence – as bathroom mirror reflections, rotating 3D models, virtual video-conferencers and health-checking, fit-looking yoga fanatics who are interested in food but don't actually eat it – manifests them as 'intensively managed' subjects within a governmental regime of gendered biopower (McRobbie 2010: 60).

While Corning and Microsoft's luminous-lustrous women are largely de-sexualized (apart from the occasional visual juxtaposition with an oddly phallic faucet), other, not altogether unrelated *i*media industries[11] are trading in less sanitized, overtly sexualized, ironically sexist mani-festations of ubiquity that scale up (or down) toward what Bates refers to as the 'Wild West' of unregulated social and online media practices (2014: 107). Bates depicts the internet as a kind of sexist 'digital enclo-sure' from which, for girls in particular, 'there is no escape,' not 'even at home, away from school and peers' (Andrejevic 2007; Bates 2014: 107). Girls find themselves 'trapped in a hostile environment' that includes Facebook, instant-messenger applications and web sites such as Snapchat 'where users send images back and forth' and anonymous question and answer sites such as Spring.me and Qooh. With little risk of discovery or adult intervention: 'Young people create a profile, including a picture, but other users can send them questions without revealing who they are. It's a recipe for disaster, or more specifically, for extreme sexualized bullying' (107). The many vile examples that girls and women have shared on the everyday sexism web site and that Bates associates, in part, with the accessibility of online pornography, include references to rape and sexual assault that are framed and normalized, more often than not as banter. Bates demonstrates how the unchecked sexism and misogyny that are increasingly prevalent on social media sites is also characteristic of music videos and computer games featuring – in foreground and as background – women in vari-ous states of undress being subjected to assault, kidnapping, rape and murder (2014). Web sites such as www.everydaysexism.com and www. feministfrequency.com are for women and girls to voice and share their experiences, objections and critiques. Bates refers to the unexpected and exponential growth in contributions to the everyday sexism site as a 'tipping point,' a point of individual and collective intervention in the form of shared storytelling. Eschewing any claim to scientific empiricism, she nevertheless maintains, somewhat convincingly, that 'the stories are corroborated by each other' (17).

DOI: 10.1057/9781137374851.0005

Feminist Frequency, established by the media critic Anita Sarkeesian, uses video series such as 'Tropes vs Women in Video Games' as critical, pedagogical tools. The second part of this particular series looks at women as background and decoration in violent games. The women-as-background trope 'is the subset of largely insignificant non-playable female characters whose sexuality or victimhood is exploited as a way to infuse edgy, gritty or racy flavoring into game worlds.' Sexualized female bodies 'are designed to function as environmental texture' and are sometimes 'created to be glorified furniture' or minimally interactive sex objects to be used and abused.'[12] As both the everyday sexism and feminist frequency sites – as well as their creators – reached a level of visibility, the cost of breaking what McRobbie refers to as the new sexual contract (voice or visibility, but not both) began to be extracted through aggressive misogynistic abuse that, for Sarkeesian, can only be euphemistically referred to as 'trolling.' In a talk for TEDx Women,[13] she speaks of insults related to kitchens and sandwiches and of how an image of her face became the central object of a multiplayer beat 'em up game. Both Bates and Sarkeesian have publicly refused to be silenced by such 'banter.'

Somewhere on a sliding scale that includes the re-traditionalization of gender roles and the evisceration of semi-naked women as background, environment or furniture; between the generic differentiation of gender and ethnicity and the aggressive, violent regression to sexist tropes of the 1950s, the *i*media industry emerges – shoving everywear women before it as shield and figleaf – with:

> Durex Fundawear, controlled via smartphone apps and designed to 'vibrate whenever your partner decides you need some sexual attention.' Tom Cheredar, on the Venture Beat news site, incorporates video stills of women everywearers and invites viewers to contribute 'slightly inappropriate and witty' comments.[14]

> Sex with Google Glass, a Google Glass app designed to 'show what your partner sees during sex.' According to Alex Hern, writing in *The Guardian*, its egalitarian promise is undermined by the focus on heterosexual couples and the fact that 'most of the images are sexualized photos of women.'[15] Sex with Google Glass will of course have to wait for the Google Glass project to be reinvented, but it was not a significant factor in the project's initial lack of success.[16]

> Samsung Galaxy Gear, a Dick Tracy/Inspector Gadget style smart watch for him that is also set to be an always on, everywearable lifelogger par excellence

for her. As Stuart Dredge points out in *The Guardian*, wearables capture more data and lifelogging, including by means of built-in cameras, will be a key feature 'in the currently hyped categories of wearable gadgets like smart watches and augmented glasses.'[17]

Microsoft's Smart Bra, to measure mood. As Dredge points out, 'lifelogging software and hardware can tell you what you did and where you did it, but apart from any related status updates, it can't tell you how you're feeling.' Now, apparently, it can. Where Microsoft emphasizes the health benefits of stress-detection and of preventing stress-related over-eating by means of integrated sensors monitoring heart and skin activity,[18] a Japanese lingerie brand, Ravijour, is promoting the idea of a chastity bra. Ideal for Cinderellas everywhere/wear, this 'remains firmly locked most of the time,' but when 'Prince Charming arrives' and the wearer's heart rate reaches a certain level – pseudo-scientifically associated with 'true love' – then 'hey-presto,' it 'dutifully bursts open with a gleeful spring, saving clumsy fumbling in the bedroom – and potentially taking his eye out in the process.'[19]

There is, of course, nothing inherently new, even about the apparently automated, algorithmic regulation, objectification, sexualization or de-sexualization of women's bodies. The fact that the smart bra works, if it works, through a combination of sensors, algorithms and apps and through the reductive metricization of affect and emotion only intensifies and extends the reach of gendered biopower. Insofar as this remains a familiar problem, it becomes further entrenched (seemingly unreachable) in contexts and environments in which feminism is undone or effectively contained not only by jokiness and illusions of equality but by internal disarticulations or, more specifically, oppositions and dialectics. In previous chapters I have considered dialectics of structure and scale, objects and relations and epistemology and ontology, but what I want to do now is look at how these hierarchical oppositions implement and organize different conceptions of time. In doing so, I'm suggesting that time is at stake in environments of ubiquity and that feminist timetelling is of strategic significance in confronting sexist configurations of women, wearables and work.

Part 2: What time is it?

Co-constituting the objects, environments and materials of imedia, ubiquitous women – who are simultaneously everywhere, everyware and everywear – emerge *as* time and *in* time. Emerging as time, women

DOI: 10.1057/9781137374851.0005

come to matter – they materialize and signify – in *potentia,* as subjects affirming the 'lively relationalities of becoming of which we are a part' (Barad 2007: 393). As time, or *potentia* – defined by Rosi Braidotti as 'the desire to become' (2006: 163) – women are construed, within aspects of feminist philosophy, as ethical subjects, open to 'encounters and minglings with other bodies, entities, beings and forces' beneath and beyond the human scale (163). The subject here is 'an intransitive process;' rather than becoming anything in particular, it evolves as a constitutive part of the generative process of life (163). Time and life are synonymous for vitalist philosophers such as Braidotti and Elizabeth Grosz for whom a-personal, or impersonal, forces of individuation and differentiation complicate identity formation, re-introducing nature into the cultural realm, matter into discourse and ontological scale into epistemological structures including those of gender, race, sexuality and class (2010; 2005). It is through a notion of time as *potentia* – rather than structured, lived or clock time – that both Braidotti and Grosz formulate an alternative, 'affirmative' feminism beyond the humanistic confines of identity politics. Without seeking to simplify or equate arguments that for me, as for many others, have been a rich and inspiring resource – including a resource for thinking about the vitality of mediated subjects and of mediation itself (Kember and Zylinska 2012) – I would suggest that this insistence on affirmation, on faithfulness or openness to *potentia* as an alternative feminism is the principal marker of oppositionalism and of dialectics wrought from a non-dialectical impulse.

Braidotti states that her 'project of affirmative politics is an attempt to relocate critical praxis within a non-necrological perspective and hence to stress vitalism and productive activity' (2010: 141). A relocation project – one that attempts to build on a poststructuralist 'hermeneutics of suspicion about the unitary subject of liberal political social and legal theory' and deal with the paradox of criticism and creativity, reason and imagination – becomes an advocacy project. Advocacy in favor of affirmation, a future or emergent politics, takes place in relation to certain assumptions about the present – 'the violence of the times' – and about previous modes of inquiry characterized by 'collective mourning and melancholia' (2010: 141–2). A social and political theory that emphasizes 'vulnerability, precarity and mortality' can, and following Braidotti's ethical imperative (rather than moral injunction) should be balanced and replaced by one that emphasizes 'relations, endurance and radical immanence' (142). It should perhaps be noted that Braidotti's affirmative alternative relies on a structured idea of time past, present and future that it otherwise eschews and that it develops from a process of deconstructive critique otherwise deemed to be too negative.

Braidotti seeks to rework the work of critique from a negative to a positive activity and from present conditions to future possibilities and alternatives. Her affirmative politics rests on a time-continuum that indexes the present on the possibility of thinking sustainable futures' and for me, it fails to move beyond dialecticism in as far as it restores rather than abandons oppositional thinking (2010: 143). Affirmation is constituted more as an opposition and alternative than as a constitutive outside of negative affects, affects that arrest movement, block becomings, and freeze us in the present time of 'pain, horror or mourning' (144). Such affects can, and for Braidotti must, be transformed, even transcended by what might ultimately be 'an act of faith' in the power of life itself – its *potentia* (144–5).

Grosz's investment in the vitality of *potentia* and in affirmative politics is more radically opposed to critique as a negative 'defensive' affect, and it is interesting, given my previous comments on the relation between critique and writing, that Grosz disavows the relation – 'the critique of texts never actually transforms texts' – in order to pursue the inventiveness and creativity that is therefore detached from and denied to the critical tradition (2005: 2). For Grosz, writing is an affirmative alternative to 'the usual critical gestures' rather than a manifestation of them. Her non-critique becomes a limited, dialectical form of critique predicated on alternative conceptions of time – structured and continuous – and, with Braidotti, on a restorative (deconstructive) strategy that turns back into an oppositional one, an advocacy of ontological perspectives as distinct from epistemological approaches and of nature as the foundation of culture. Ironically, it is in the attempt to distance herself from the inevitability, the restorativity of oppositional structures and their 'temporary' (rather than temporal) breaches that Grosz establishes this foundationalism. Derrida acknowledges 'the tenacity of the binary structure' even as he proposes 'a tripartite strategy' for shaking it up and provisionally unhinging it (6). Where he offers 'reversal, displacement, and the creation of a new or third term' that has proven, Grosz acknowledges, important to a variety of political projects including feminism, she declares her preference for a vitalist philosophical tradition informed

DOI: 10.1057/9781137374851.0005

by Bergson and Deleuze in particular. While this enables her to advocate a 'reconstituted ontology' premised on time as life, on movement rather than stasis and on the becoming of politics as well as of nature, it also leads her to propose that a reconstituted ontology is the 'ground' or foundation of knowledge practices, 'the very terrain out of which the dominant term – culture, mind, representation, self, subject, intelligence, the determinable, the solid – emerges ...' (5,7).

Alternative concepts of time are inherent in the vitalist tradition. For Bergson, structured, lived or clock time is spatialized time, a necessary expedient and instrumental cut into and against continuity and the in(di)visibility of creative evolution (Kember and Zylinska 2012). Structured time is atomized, organized, mechanistic and finalistic: oriented to ideologies of progress, performance, productivity and potential. Potential and *potentia* constitute the temporal axis of contemporary feminism, one, I suggest, that need not and – in order to allow for a feminist politics that is both emergent and contingent – should not be construed as oppositional. I am arguing that potential and *potentia* are co-constituted and that it is not possible to choose between them on the basis of a feminist ethics or politics. If choice or substitution is the basis of a limited feminist critique and an impoverished feminist political theory, then a more effective approach to understanding and addressing the current ubiquity of women might be based on an ongoing attempt to cut again and against or to intervene creatively and critically in the cuts in *potentia* that industry as form of governance makes by micromanaging women in time. How might Jennifer's bathroom mirror schedule and Ayla's car window countdown be countered, reversed and refigured? Is there a time machine for women? If time is the mechanism and glass is the material for openings and closures, then what might a feminist praxis make of and with them? I have argued elsewhere that a recourse to the notion of sustainability will not cut it, precisely because it is predicated on limits that are constantly and increasingly being over-ridden (Kember and Zylinska 2012). Carol Stabile[20] and Rosalind Gill have both made powerful arguments about the unsustainability of the work that is being cut out for women in a neoliberal economy (Gill and Pratt 2008; Gill 2010). If, with Laura Bates and Anita Sarkeesian women find that their capacity for endurance (a capacity that is purely ethical for Braidotti) is demarcated if not delimited by its conditions of im/possibility,[21] then feminism should focus its critical-creative strategic intent on the un/endurability of ubiquity .

DOI: 10.1057/9781137374851.0005

Notes

1   For Adam Greenfield, everyware is computing that is 'ever more pervasive, ever harder to perceive' (2006: 9).

2   What follows here is a kind of re-storyboarding, a re-telling of Corning's tales of A Day Made of Glass. These promotional videos are, I contend, potent stories, not just in their virality (YouTube figures in the tens of millions) but in their capacity to enchant, to cast a spell, to suspend disbelief in the wonder of glass things that just are, or very soon will be. One such object, or becoming-object is Jennifer. Her husband or partner is not named in Corning's tale – a tale that leaves him out, that lets him off and allows him, in his transparency – a transparency that mirrors that of glass itself – to just be. Jeremy is only named as Jennifer's alliterative other in my re-telling of Corning's tale – one that is not quite as ready to let him off.

3   https://www.youtube.com/watch?v=6Cf7IL_eZ38

4   https://www.youtube.com/watch?v=jZkHpNnXLBo

5   As of 2012, and the cost of switchable glass remains very high as of 2014.

6   https://www.youtube.com/watch?v=X-GXO_urMow

7   https://www.youtube.com/watch?v=phH2TdCyWbs

8   https://www.youtube.com/watch?v=a6cNdhOKwio

9   https://www.youtube.com/watch?v=DoCCO3GKqWY

10 Lecture for the Goldsmiths Feminist Research Centre, Wednesday January 21, 2015.

11 I'm mainly referring to start-up companies here, but as Stuart Dredge writes in *The Observer*, 'big tech companies are sniffing around' them ('10 things you need to know – about lifelogging,' *The Observer*, Sunday February 9, 2014.

12 www.feministfrequency.com

13 https://www.youtube.com/watch?v=GZAxwsg9J9Q

14 Tom Cheredar 'Durex creates vibrating underwear you can control via Smartphone apps,' VB news, April 19, 2013 www.venturebeat.com

15 Alex Hern 'How to make sex 'more awesome' using Google Glass,' www.theguardian.com, Tuesday January 21, 2014.

16 This, as Chris Matyszczyk reports, had more to do with it being unfashionable, nerdy and somewhat extraneous to the communicative requirements of a public still enamoured of smartphones http://www.cnet.com/uk/news/dear-google-glass-bye-for-now-and-maybe-forever/

17 Stuart Dredge http://www.theguardian.com/technology/2014/feb/12/10-things-to-know-about-lifelogging

18 http://www.bbc.co.uk/news/technology-25197917

19 Oliver Wainwright 'The chastity bra that only opens when you find true love – or so they say,' http://www.theguardian.com/artanddesign/architecture-design-blog/2014/jan/31/chastity-smart-bra-opens-true-love-japan-lingerie

DOI: 10.1057/9781137374851.0005

20   Carol A. Stabile 'Magic Vaginas, The End of Men, and Working Like a Dog',
     paper presented at Goldsmiths, University of London Annual Gender Event:
     Feminism and Intimacy in Cold, Neoliberal Times (2013).
21   For Sarkeesian, these conditions have included death threats that have led
     her to cancel a university talk: www.theguardian.com/technology/2014/
     oct/15/anita-sarkeesian-feminist-games-critic-cancels-talk

DOI: 10.1057/9781137374851.0005

4

interlude 1: Excerpt from *A Day in the Life of Janet Smart*

Abstract: A Day in the Life of Janet Smart *is a novel in progress that temporarily occupies Philip K. Dick's* Minority Report *and parodies the extent to which its predictive-responsive i-enabled glassworlds inform corporate and computational visions of our smart futures. Skeptical about what these futures have already delivered in terms of intelligent labor-saving devices such as care robots and self-driving cars, Janet starts her day with a serious case of jet lag, juggles her job, her professional, personal, human, non-human and casual relationships and then has to figure out how to stop her daughter living the same unsustainable life on a whole other unsustaining planet.*

Kember, Sarah. *iMedia: The Gendering of Objects, Environments and Smart Materials.* Basingstoke: Palgrave Macmillan, 2016. DOI: 10.1057/9781137374851.0006.

07.33 London; 15.33 Seoul

Janet Smart was appreciating the quiet companionship of her car. Colin, as she sometimes called him, had greeted her by name at the airport car park, asked about her trip, registered her fatigue with the seat (heavy) and door (weak) sensors and tried his best to shut up. With Colin in control, Janet was able to check her makeup in the driver's-side pop-up screen that displayed the time, time to destination and world clocks at recent and related locations. 07.33:46:24 and counting. Seoul, 15.33; Beijing, 14.33; Kuala Lumpur, 14.33. Janet had been flying for 26 hours and 45 minutes including the stop in Tokyo. Her employer funded business class, but this was not always the same as first class, so she had the upright equivalent of bedhead, and her pinched skin and shadowed eyes showed signs of mild dehydration. Still, she thought, as she retouched her lipstick and buffed her cheeks, not bad for someone staring down the barrel of middle age. The menopause was still a few years off and that proto-wattle could be dealt with, if not by surgery, then by those exercises she'd looked up. Opening her mouth as wide as she could, she bared all of her teeth and stretched the sinews and muscles in her neck. Then she made a tight 'o' with her lips. That's what got the job done, apparently. She was supposed to repeat this rictus and rectum regime as she thought of it at least fifty times every day. Just half of that made her face hurt. Janet ignored the rude young man in the car that overtook her and got as far as thirty when the purple light on Colin's dashboard indicated that her 7-year-old daughter was calling.

As Chief Operator of the capital's Pre-Consumer Agency, Janet was not often at home. She was returning from an international conference in Korea where pre-consumption had been widespread for over a decade. Their technologies were state-of-the-art and their enforcement practices – Janet preferred to think of it as consumer consolidation – were hyperefficient. She was there to learn from them, but what she'd come away with was a stronger sense of doubt. Her doubt did not concern the agencies themselves. Consumerism had been opposed in the past, she knew, and mostly by middle-class hippies and hypocrites who were content to deny others the stuff they took for granted. Real concerns, about natural resources and global warming had been lost, overtaken in a global fight between the haves and have-nots that had turned the market into a monolith that in turn had generated, for its own survival, the environment neutral devices they had today. It was certainly a

DOI: 10.1057/9781137374851.0006

twisted world in which insurance companies became the main funders of sustainable technology, but it was the world that Janet lived in and was therefore obliged to deal with. She remembered when that superstorm flattened half of Manhattan. Her partner at the time, the man who would become the father of her child, called her from work and told her to look at the news. She was busy and was short with him, but he insisted. She watched the deluge and the carnage on a loop, spellbound and horrified as a growing wall of water surged through Brooklyn, Queens and the Financial District. There was a value put on that, in lives, but also in dollars. Now, and this was her issue, she could no longer tell whether technologies existed to save lives or to save dollars. Who, or what exactly were they sustaining? The one thing she was sure of was that, present company apart, they were not there merely for her own comfort, care or convenience.

> 'Mum, I feel sick.' Janet always felt weird when her face morphed in to Jennifer's. It was the lack of correspondence if anything, the sense of a life to be lived differently, perhaps even tangentially. Janet didn't mind that they were dissimilar. To the contrary, she loved her daughter's willfulness, her sense of self, but she worried that her absence was the cause of it and of whatever effects might ensue.
>
> 'I can see that. Has Zachary taken your temperature? You haven't been experimenting again have you?'
>
> 'Yes he has and I told you that. We're making my wallpaper. I told you yesterday.'
>
> 'That was the middle of the night for me darling: your night, anyway. What did Zachary say?'
>
> 'A hundred and four and guess what? He wanted to get me a hot water bottle.'
> 'Stupid machine!'
>
> 'Mum! Don't call him that. It was funny. I couldn't sleep so he's been helping me.'

Jennifer had known Zachary her whole life. He was her care robot and Janet struggled with that, trying but invariably failing to deal with her envy and resentment. He was not a humanoid robot but that didn't seem to make any difference. Jennifer loved him and it was that simple. She understood clearly enough, thank you very much, that he was not human. What kind of slow tool adult would think she thought that? He ran on wheels, for one thing, and even the new robots with legs made funny noises when they walked, like wheezing and sighing, only not from

DOI: 10.1057/9781137374851.0006

their lungs because they didn't have any. He couldn't do things like sport but so what? She wasn't into that kind of thing anyway. Jennifer had resolutely refused to have Zachary replaced. Ever. On account of his phase two mechanics, Janet had indeed ruled out sport. His throwing arm, for example, was pneumatic, and she was pretty sure he wasn't waterproof. His reading voice had destroyed any nascent love of literature, but Zachary had instilled in Jennifer a love of experimental science, and he was currently nurturing her ambition to be a biodesigner. Together, they had been growing her bedroom wallpaper out of mosses, lichens and small plants. They'd been working on their own species with tiny, delicate orangey-red flowers that would protrude through holes in the double-layered photovoltaic glass containing capsules of nutrients activated by rainwater filtered down from the roof. Maybe Jennifer's sickly hue was just a reflection of all that green stuff on the worktop in front of her.

> 'Have you been eating properly? I've told him a hundred times that Froot Loops don't have fruit in them.'
>
> 'He knows.'
>
> 'Not really. You need to go back to bed. Try and sleep. I'll message him and check that he's told Mrs. Thompson at school.'
>
> 'Mrs Thompson will be notified at 08.30. Good morning Janet. How was your trip?'
>
> 'Why haven't you called the doctor, Zachary?'
>
> 'Jennifer vomited into the Massashitsu Mark 1 automatic toilet at 06.42 and I will receive the full analysis in 29 minutes from now. Her mirror reading is not conclusive. However, I have reviewed the symptoms of mumps, chickenpox and three other juvenile afflictions and can report no measurable statistical correspondence.'
>
> 'Great, so it's probably the flu.'

Janet flicked away her personal screen and changed from self-drive to manual. As she did so, Colin's windscreen displayed the route and remaining journey time.

The all-screen Augmented Reality Flags and Edifice Outlines had been set to Major Landmarks and Roadblocks Only, but it kept defaulting to Retail, Fast Food and Coffee Outlets. For a moment, Janet could hardly see where she was going.

> 'AR off!' She ordered, as the amber light flashed to indicate an external call. Red was for her mother.

DOI: 10.1057/9781137374851.0006

'Yes, hello, who is this?' she asked; failing to recognize the voice of the man she'd slept with on her third night in Seoul. She was bored with networking by then and he was young, fit and if the translation provided by her watch was anything to go by, really funny. He was also, unfortunately, smitten.

'You are a bountiful woman Janice,' she heard, moments before he appeared to mime the words. 'I am think move to UK.'

'I am think not,' she muttered, cutting him off and wondering what Colin's voice translator was up to. This guy, what's-his-name, couldn't speak a word of English, so what was with the caricature? Was Colin starting to think that he was funny too?

'Bring up home AmI.'

As she waited for her house Ambient Intelligence System to respond, the faint, blurred film of an unromantic but energetic encounter played in her mind. She had obviously misread his ardor for lust since that is all she had signed up for that night. She remembered them both laughing as they slid down the gap between the twin luxury double beds and was just retrieving the sensation of a taut buttock in her hand when the familiar intro signaled the readiness of her ambient intelligence.

Due to the well-documented problems of voice recognition, it had been necessary for Janet and Jennifer to invent names for the various household systems. If they couldn't figure out what you were saying, they might at least know which of them you were saying it to. At the time though, Janet had not thought through the consequences of deferring the naming process to her then-5-year-old.

'Fart.'

Nothing.

'Fart!'

'I heard you say: fart. Is this correct?'

'Yes, brilliant. Now put me through to the fridge, please.'

'I'm sorry. I didn't hear that. Could you repeat.'

'Fart.'

'Connecting.'

'Good morning Janet, how was your trip?' Bum asked, cheerfully. Pausing for an answer but not getting one, it proceeded to ask about her bathing requirements for the day. Janet would have killed for a nice hot bath. Briefly, she thought of filling one and pushing Zachary in. Then she thought of Jennifer. She ordered a bubble bath for her, one-third full, for 18.30 and then managed eventually to get back to the fridge in order to make sure there was enough milk, juice and tomato soup. That left Crayon the curtains, Lolly the lighting and last but not least, the Heebie Jeebies.

'Raise the heating in bedroom two by ten degrees,' she instructed, hoping to sweat Jennifer's virus out.

During her quality time with the house, the amber light on Colin's dashboard had not stopped flashing. Now it began to alternate with a red one.

'Jesus,' she hissed. 'Mum. How are you doing?'

'Well dear, I thought I'd better get in touch as I haven't heard from you for a few weeks. Hello. Can you hear me?'

Janet did consider de-selecting the red channel, but, not wanting to be outdone by a car, she did something worse.

'Can you hear me dear?'

'Oh yes. I hear. You are bountiful, bountiful woman and I never stop thinking you.'

'Who is this?'

'I am Li-Kuo. You are not Janice Smart?'

'Yes. I'm Janice Smart. Have we met?'

Leaving him with that thought, Janet slipped out of the conference call and, still 8 minutes, 26 seconds from the agency, decided to check in quickly with Marvin.

'Polly has something. Something big. But she's playing up again. I've got a couple of names but she won't give me anything else. Tell her, will you?'

'I'll be there soon. Please try and get along 'til then. And the trip went well, thank you for asking.'

Jennifer's line was flashing again.

'Mum, what have you done to the heating? It's boiling. Zachary already turned it up.'

'I'm sorry. He didn't tell me that. I'll turn it down again.'

'No. It's okay. Don't. The plants like it. There's this one, it's skinny and needed a stick to grow up. Its leaves were all tight shut, but now they're opening and they're all sticky. Zachary said we should feed it. I have to go and get some flies and beetles and stuff from the air filters. They're high up so I need the ladder. Its in the basement I think and you had the key in your room because, I dunno, you've got a secret down there or something.'

'Drugs. Alcohol. Dismembered bits of robot.' Janet said, under her breath.

'What? Anyway, what should we do?'

DOI: 10.1057/9781137374851.0006

'Nothing. I'll find the key for you tonight, although I'm afraid we might not have a ladder anymore. If you're good and get some rest, I'll bring you back a Mackie-D for your triffid. Ok?'

'Ok. What's a triffid?'

'Nevermind. Gotta go. Love you lots.'

As Janet pulled in to her parking slot, she had a sneaky listen to the ongoing conversation between her mother and her one-night stand. They seemed to be doing fine. She set Colin's communication channel to auto and put him on charge. Her gait had been measured and classified (weary) by the time she got to the lift, which, apart from its obligatory greeting, restricted itself to announcements and notifications only until she reached the sixth floor. The entrance to the Pre-Consumer Agency was not very subtle, but it was mood-sensitive. The floor-to-ceiling transparent glass screens profiled key personnel and ran interviews and endorsements from politicians and celebrity consumers. The usual cacophony of colors and voices was suddenly muted, and Janet got past reception while Andre, the receptionist, was having a not-very- private discussion with his boyfriend in Germany. Inside the agency suite (only one of them called it a lab) she gave Marvin a warning glance, placed her mobile on her desk and was presented with the day's work in childlike, supersized text and gratuitous holographics. First on the list was case 4,510,342; nearly half of the population of London, as the recent public relations campaign had trumpeted. Paul Briddel and Mary Sommer were, according to the confidential Ident from Polly, procurement officers for IASA, the International Aeronautics and Space Agency.

Earlier at the agency

'Macaroons!' exclaimed Polly the Pre-Con, her expression setting to anxious-shocked. The intermediate states were fitted as standard to the Emos, but Polly was a first-edition, and an upgrade was overdue.

'Multi-colored. In a box. Two layers. She's ... on a diet!'

'Name?' Marvin asked, without lifting his head from his flexiglass ebook. He was reading *The Time Machine* for the forty-third time. To him, it was more than a classic. It was a document, a manifesto for future travel that had been wrongly classified as fiction. Reluctantly, he selected menu and bookmarked page eighty-two, the part where the sun had grown red and large and the Earth was heading toward entropy. Polly hadn't finished talking, but Marvin

DOI: 10.1057/9781137374851.0006

had already decided that, on this occasion at least, she didn't have anything interesting to say.

'Zap bag containing two skinny fit tees and new season retro cords. Tag reads size twelve. She didn't try them on. They're not…going to fit! She only has 24 hours to exchange…'

'Name!' Marvin barked. It wasn't that he didn't believe in the Emos and their superior predictive abilities. He had done courses in affective computing and data analytics at Imperial and had been drawn to work, for less money than he might have secured elsewhere, with the ground-breaking Pre-Cons, the mutant, emergent artificial life forms that were the unforeseen, almost miraculous outcome of massive data input, embodied emotional intelligence and good old-fashioned neural networks. Polly was a geek's Goddess but what Marvin couldn't stand was the breadth, the veritable Catholicism of her event filter. She might as well not have one. What women bought to eat or wear was of no earthly consequence. Correction: it was of only earthy consequence. Is that what the Pre-Consumer Agency was for? That's not what they'd led him to believe at interview. Then, he'd been offered a vision of advanced technologies and ambient intelligent environments that would first cover the entire globe and then enable select groups to ship out to other planets. It wasn't going to happen tomorrow, of course, but it certainly clarified their objective: namely, to ensure that what the Pre-Con prophesied always came true. The marketers did what marketers do, but there was little point in people dreaming of their speech-enabled self-drive car if they didn't actually buy one, and if they didn't buy one they were hampering, say, rocket development and robotics. The agencies owned the banks, so loans were no obstacle, and it no longer mattered if the borrower defaulted because the agencies – there were four in England, but policy was led by London and the South East – could always compensate themselves. It was a vision and a goal that Marvin believed in and dedicated himself to. This is why he became frustrated when he felt that Polly's talents and, indeed, time itself were being wasted.

The name Sal Evans appeared on Marvin's desk screen and, as if to spite him, Polly projected Sal's photo identification, complete with biometric and biographic information onto the display screens that dominated the room. Marvin pushed his chair back, stood slowly and walked toward Sal's face, becoming, by degrees, more erect, more human. He really liked this bit, regardless of the potential purchaser. His gestures were direct, expansive and emphatic, like a conductor isolating members of his orchestra, holding some back, bringing others forward and lifting the whole for a crescendo of sound. Sal's data, though of no importance to Marvin, scattered and combined, pulsing as he scanned it. With an

DOI: 10.1057/9781137374851.0006

authority and dexterity he lacked outside of his lab, Marvin drew Sal's vital statistics, body mass index and recent medical history toward him so that they stood out from the transparent screen, and with his other hand he flung away the animated images of her youth, her memories of fashion and desirability that had encroached on her present and, like a subliminal advert, generated her intention to shop.

> 'Should have gone to goggle-savers', Marvin said as he brought up Sal's half-scale hologram and spun it around faster and faster with his index finger.
>
> 'Stop!' Polly intervened, showing Marvin the pre-scene of Sal approaching the Pay Here desk.
>
> 'Dispatch digital assistant', Marvin ordered, having already turned around and headed back to his desk and his book. He did not see the smiling assistant approach Sal and recognize her as one of Zap's original fashion designers. Marvin hadn't bothered to read her bio or, for that matter, her credit rating. There had been problems with the digital shop assistants, particularly with their tactful-truthful balance, but the job for the DA3 model was made easy by the fact that Sal was late for a board meeting, didn't have time to search her bag for her reading goggles and had simply picked up the wrong sizes. The macaroons were for sharing. She was proud of her work, proud enough to wear it, and grateful for her assistant's advice as well as the way she gave it. As Sal faded from Marvin's screen, two new names appeared, and this time, Polly knew very well that he'd be interested, so she waited until Janet arrived, 27 minutes and 41 seconds from now.

Marvin's mission

Marvin should have been transfixed by what he was seeing. Instead, he fidgeted, folding and unfolding his arms, pacing this way and that, gazing at the screens and then at his feet. He had an annoying habit – well, several – as far as Janet was concerned, and one of them was chewing the skin at the corner of his nails. He did this systematically, both corners, working from thumb to little finger on his right hand and then his left. He was doing it now. The IASA officers were inside a show pod, a ME Bio-home living space designed and developed for the delayed 2024 one-way mission to inhabit Mars. In contrast to Marvin, Polly was enthralled-ecstatic.

> 'It … *breathes* for them!'
>
> 'Ok, keep your levels down', Marvin snapped. 'She can see that for herself.'

DOI: 10.1057/9781137374851.0006

Janet looked at the grassed glass interior walls and saw, with immediate impact, what this was and what it might mean for her family. That this pod, and, God forbid, its intended destination represented a possible future for Jennifer is not something she thought or could, for the sake of professionalism or the always urgent, decision-making demands of her job, push to the back of her mind. It was delivered to her by an invisible fist, an expert, black-belt punch to the solar plexus. It was all she could do not to fold over and fall to her knees. Masking her breathlessness as much as she could, she dispatched the digital assistants as standard, a male for the male purchaser and a female for the female. Paul Briddel was white, middle-aged and balding. His fellow officer was a young black woman with sleek, straightened hair cut into an angled bob with a low fringe skimming across the top of where her eyebrows used to be. A heavily drawn line arched upwards from her brow giving her a look of permanent cynicism without any need to mobilize the rest of her face. The mirroring process was automatic and, in Janet's experience, invariably effective. Mary Sommer did not look convinced but Paul seemed pleased to see his digital doppelganger. It was important that these remained non-threatening even while they flattered their human counterparts. The trick was to give potential consumers an image of themselves that was idealized in the right way: not necessarily younger or better-looking but more decisive and assured. This was easy enough to do; after all, bits of software, however sophisticated, have no experience and therefore no basis for doubt or indecision. Paul's assistant introduced himself in a voice that was the equivalent of a knuckle-crunching hand-shake. This combined well with his height and his hard-earned, held-in waistline paid for by many hours at the gym. Paul took a step closer and leaned in as his digital self explained how the pod worked and outlined the principles of biodesign.

Although they were now in real-time, Janet, breathing shallowly to avoid the physical pain in her stomach, paused the scene and parked it on the review screen. She would have to look at it later. By making a vertical cut with the edge of her hand, bringing her palms together and separating them, she created two parallel events from the single stream that was coming through Polly. Paul was now being shown the eye-tracker Face Recognition Television using technology of the sort, Janet reflected, that still got people falsely arrested at airports, especially if they were not white, middle-aged and balding. The FRTV knew who was watching it and exactly what they were looking at. It sought to match

DOI: 10.1057/9781137374851.0006

that information with suitable viewing content, meaning, Janet supposed, there would be no real need for porn channels anymore. While Paul kicked back in the ReAction armchair that was synched to the television, Mary was taken on a tour of the kitchen. As a career woman who ate business lunches or not at all, she was almost taken aback when the kitchen worktop asked her if she'd like some help with her baking. At this point, her assistant pointed emphatically at a bag of flour and then at a hot spot on the gleaming surface where the flour was meant to go. Mary duly moved the flour, holding it as she might a newborn baby, afraid that she might drop it or that some of it might spill. Placing it down carefully on the now glowing spot, she was offered, verbally and in text form, a number of artisan bread recipes, each with its own synthesized smell and three-dimensional image that rotated, teasingly in the sterile air. Unimpressed, Mary asked where all the wheat would come from on Mars. In response, she was invited to touch a transparent interactive markerboard that popped up from the floor and revealed, in an instant, the plans for a giant biosphere. The biosphere would support the first twelve inhabitants of Mars and the six living pods they would share, one couple per pod including four heterosexual couples (three white, one black), two Asian lesbians and a couple of male Astrophysicists who had been best friends since school. Janet found such tokenism offensive. It was a feature of the tech industry adverts and now IASA were going to make the industry's limited, generic vision of a mixed society the reality of life on another planet. It was a wholly disingenuous claim to equality and Mary could hardly fail to catch that too. She was hoping to get on with the tour and could hear Paul talking to what she assumed was the pod's companion robot. While they discussed the chosen location for the Martian community, Mary, who thought that one patch of red dust was probably as good as another, was shown how to operate a virtually invisible glass hob and found out that if she put her medicine bottle in a particular place, the sensors in the ceiling would read the tag on the label and remind her when to take her pills.

'We're losing her. We're going to have to go ambient.' Marvin sounded panicky.

Janet did not approve of ambient forms of consumer consolidation, preferring, if at all possible, to be more upfront.

'Come on Janet,' Marvin insisted. 'We have to. You know that. This is about as big as it gets and if we don't secure the sale we're done, we're nothing. Game over.'

DOI: 10.1057/9781137374851.0006

Janet still looked doubtful.

'For Christ's sake. What is it with you? What are you waiting for?' As Marvin was clearly in the process of becoming hysterical, Janet decided to let his insubordination pass, for now.

'THEY ARE GOING TO M-A-R-S,' he spelled out, with his eyes as well as his words in capitals. 'Twelve lucky bastards will get on a rocket, take the ride of their lives and one year later they will be settlers in a new world. They will be out of here.'

'You envy them?'

'Shit yes, I envy them. I would give anything to be one of them, but I didn't make the grade.'

For a moment there was silence and Polly, who had twenty-five per cent of her attention awareness circuit channeled through the room looked, as if on Marvin's behalf, shifty-embarrassed. Marvin was embarrassed less at this revelation to his boss and more by the fact of his failure to be selected for the mission. This put him in the same category as all the other rejects, some of whom, no doubt, had been serious candidates like himself, but the rest, the majority, the ones whose thumbnail bios were broadcast all over the news, were nutters, eccentrics and outcasts, people whose life on earth was such that a one-way trip to a planet with no air, life, water supply or food seemed like a really good idea: losers, in a word.

'It doesn't matter anyway,' Marvin lied. 'I'm going to apply for the Europa mission as soon as it comes up, and I've been told my chances are good.'

'Okay then.' Janet thought it best to pause there and besides, she could see from Polly's data stream that Mary – who was now in front of an augmented reality bedroom mirror that displayed atmospheric conditions, assessed the suitability of her clothing and gave her a to-do list for the day – was still wavering. Marvin the wanna-be Martian might just be right.

Ambient enforcement, as it was commonly known in the agency, was justified to the general public as a necessary means to an end. That end, Global Democratic Capitalism (GDC), was, since the third and final Arab Spring, universally held to be a social good and the only viable political system. People had come to embrace their citizenry not just as actual but also as potential consumers. Indeed, this potentiality was all-important, adding depth to what had once been considered shallow and ensuring that the most ordinary, even wretched of lives held meaning. Everybody, including Marvin, was somewhere on a scale of

DOI: 10.1057/9781137374851.0006

wanna-be and wanna-buy. While it was possible to function at either end of the scale, to pursue success or contentment, there was always the risk that you might not get there, and the purpose of a culture that had calculated risk was to eliminate it. Happiness then, was a risk-free mid zone in which being and buying were synonymous, guaranteed and publicly sanctioned. The only flaw in this otherwise hermetic scheme was that the likes of Marvin were not much interested in happiness while many sections of the global populace were no nearer to achieving it than they'd ever been.

Marvin watched Janet watching Mary and mentally rehearsed his exit strategy: get promoted from Senior to Chief Operator (Janet couldn't possibly last much longer – look at her, she's clapped out and her decision-making abilities, once impressive, are now shot) or in worst-case scenario (secret plan to go for second-wave Mars migration fails and/or cryogenesis for Europa mission declined due to hereditary vascular problems and likelihood of pulmonary embolism once defrosted), agency Director. This would mean being grounded for an entire lifetime, but power would be guaranteed. Once policy-making experience is established, consider a move to the GDC and be the person who finally introduces the new and much needed eco-eco equilibrium laws that not only recognize but maintain the alignment of ecology and economics, safeguarding Gaia 5.0 and imposing nothing more onerous than a simple per capita spending quota, differentiated, of course, by consumer category. It wasn't good enough for GDC leaders to throw their hands up and say that such quotas were impossible to administer. Such calculations might be beyond them, but that's what algorithms are for. Algorithms are putting numbers on each and every one of us all day and night. It would be nothing to amalgamate them into a single value. He could probably write the software now. He might as well, as Janet is still standing there, still watching Mary who is trying to get out of the bedroom and, in all probability, out of this deal altogether. The good thing about not having quotas, Marvin admitted to himself, was that all of the inhabitants of the developed world had submitted to these invisible, subliminal forms of intervention, hidden forms of emo-marketing that worked like magic even if they were hard to control.

Going ambient meant switching from human-machine to machine-machine communication. It relied on networked intelligent systems – from the Emos like Polly all the way down to cars and fridges – and on the existence of a decentralized hive mind, a brain without a body or an identity, at least as a human would understand it. Marvin understood it, or he thought he did. A complex distributed system had emergent effects, properties that could not be predicted – except in their complexity. One of these effects was an uncanny ability to know what to show, to which pre-consumer and when. It could be a sound, an image, a smell or a feeling, and often it was a combined sensory evocation that proved more persuasive than an appeal to reason ever could. For Marvin, there was nothing immoral about this, since political consensus had already been achieved and the citizens over whom he already had some jurisdiction had tacitly agreed to what, in effect, was a temporary brain bypass.

Having watched Mary for a little under five minutes, Janet gave the order. They stood back and watched as Mary, currently glaring into the eye-scanner on the pod's main door, was bombarded with messages neither she nor anyone else was able to decode. What Janet and Marvin witnessed was the physical and mental result of sensory overload, and to Janet at least, this looked too much like distress. She winced as Mary appeared to struggle, holding her hand to her flawless forehead and closing her large, silver shadowed eyes as, Janet imagined, her deepest fears and desires were being probed, given shape and form, turned into a story that involved a heroic mission to Mars. Ambient enforcement was intense but mercifully quick. Mary seemed to be coming out of her momentary trance when Polly, who was normally silent, reduced from being Polly to part of a collective intelligence, surfaced unexpectedly:

'Do you see?'

This was clearly a rhetorical question. Even as Polly spoke, she appeared to drift away again, as if she had sunk into her own trance-like state. Her humanoid frame, which was usually reclined, twitched then jerked upright. 'Do you – see!'

Marvin's frustration hit a peak. He actually hated himself when he shouted at Polly, not because she was the closest thing he had to a companion but because she was the closest thing there was, so far, to post-human life. In a sense, this also made her too real for him, especially when she became, in his view, distracted from her higher-end tasks. If he, as a mere carbon-based entity had eliminated trivia from his mind, then, with access to an infinity of intelligences that he could only dream of...

DOI: 10.1057/9781137374851.0006

'No, we don't see! You know damned well we don't see!' he yelled. 'Why don't you just tell us, since you're the one who's in there making it all happen? Well? What's going on? Please God make it something that matters. Hello? Earth to Polly. What is it?'

When Polly turned to look at him, she didn't seem to know.

There were nearly fifty other cases that morning, none of which put Marvin in a better mood. In fact, their daily enforcement rate had been dropping, a sign, as Janet and her superiors at the agency had agreed, of their success. Nobody ever needed enforcing twice. Janet sent Marvin for his regular break at 10:30, and, as usual, this coincided with the first of Polly's daily naps. While Polly auto-selected dormant mode, Janet decided to review the material on biodesign. Many homes, though not her own, had some biological components – moss tables, bioluminescent chairs – but she had thought them largely decorative. The Mars mission relied on the development of functional biomaterials of the sort that would fascinate her daughter, who would be 26 by that time and potentially very qualified. There was no way Janet could contemplate letting Jennifer take a one-way trip to a dead planet in the hope of making it livable for the privileged few. Not that she could stop her, and not that they would be privileged at all. It was bad enough having to talk to your house without having to weed it and feed it too. And how do you mow a grass wall anyway? They would have a device for that, of course, a gravity defying mowbot, and that would be as needy and demanding as the rest: What time should I mow, how close should I mow, would you like stripes in your wall-lawn and oops, sorry, I just trimmed your cat. Well it should have stuck to its scratch post. Shall I order another one? What color? Pure-bred or spliced? Cogs are popular right now. Litter-trained but less high-minded. Better still why not try a myRobocat? They are programmed to learn as well as speak, and unlike the Cogs, they have completely authentic species characteristics including superiority and attitude. You can have no doubt that you humans will finally discover exactly what cats say to you when you offer them food that is not to their satisfaction. Marvin is right to be pissed, Janet thought, because this is what it all boils down to. Earlier this century, there was a backlash against dataveillance and people's homes were slowly made private again but, in another nod to Marvin, she acknowledged that these same homes were very like time machines. The further they went forward in time, the further they went back, and

DOI: 10.1057/9781137374851.0006

we're talking about domestic servitude here, care work without end and no, it wasn't that she was paranoid and hated technology, she just wanted it to grow up, stop asking her questions, stop pretending to be something it was not. If the home of the future was, in effect, a lie, the biological home would be a much more dangerous one. Janet had never felt so worn out.

DOI: 10.1057/9781137374851.0006

5

iMedia Manifesto Part II: Tell a Her Story: On Writing as Queer Feminist Praxis

Abstract: *What can writing still do?[1] For the proponents of post-politics and post-humanism (in which post means after), writing's time – the time of and for writing – is up. Is writing no more than a protectionist racket for the impotent I, the dreamt of and dreamt up subject of political theory now substituted by a world of objects, environments and materials as things-in-themselves? The dismissal of writing and/as critique works towards the realization of an unmediated i (meaning immediate, intuitive, intelligent ...) world in which an undifferentiated citizen-consumer gestures and senses but is effectively made mute. But who speaks and writes this world that dismisses its own wording and disavows itself as a futuristic fantasy and science-fiction his story? The question of who writes (and to what end) partly answers the question of why write (and to what end). Writing works not as the realization of fantasy worlds that require a suspension of disbelief, but rather toward the unmaking and remaking of them. Its role is, as it always was, both de- and reconstructive: neither negative nor affirmative. As a queer feminist praxis, writing reverses and displaces the gendered hierarchical dualism of naturalized entities like subjects and objects (Haraway 1991). Its strategic value is that of reinvention rather than mere substitution (her story never simply replaces his) and it constitutes, as an action, in its present participle, a way out of the dialecticism that precludes the possibility of doing iworlds differently.*

Keywords: critique; dialecticism; writing

Kember, Sarah. *iMedia: The Gendering of Objects, Environments and Smart Materials.* Basingstoke: Palgrave Macmillan, 2016. DOI: 10.1057/9781137374851.0007.

Writing manifestos

> Write, let no one hold you back, let nothing stop you …
>
> (Cixous 1985: 247)

Manifestos are inherently writerly. They are, traditionally, exhortatory, polemical, unapologetically from somewhere rather than nowhere and everywhere and antithetical if not opposed to being right. I see rightness as a marker, even a disavowed manifesto for neoliberalism. According to this rationality, in which moralism displaces antagonism, it is right to be right, to be grounded in a reality that is, in the first and last instance, economic (Brown 2005; Mouffe 2013). If rightness, alongside fitness and productivity is the default mode of what Brown terms 'homo oeconomi-cus' – the de-politicized, de-differentiated and unmediated neoliberal subject that is (thereby) subject to remediation through the terms of gender[2], race, class and sexuality – to what extent is the alternative vision to which Brown refers (of which she speaks-writes) contingent on the tension between rightness and writerliness (40)? This tension, a genera-tor of political potential, is evident in contemporary manifestos such as:

Lindsay German 'A Feminist Manifesto for the 21st Century' (2010)

Guerilla Girls 'School of Art Institute for Chicago Commencement Address' (2010)

Lucia Tkacova and Aneta Mona Chisa 'Manifesto of Futurist Woman (Let's Conclude)' (2008)

Martine Syms 'The Mundane Afro-futurist Manifesto' (2013)[3]

German's claims concerning the reality of globalization and neoliber-alism – said to 'have had a profound effect on the lives of millions of women' and to have created 'new forms and manifestations of women's oppression' – are listed 1–12. Listing, as an anti-literary tradition, is the constitutive outside of the sort of narrative employed by the Guerilla Girls. Their manifesto, performed by an anonymous speaker in a gorilla mask, consists of an autobiographical review of a poster campaign 'about the state of women artists in the New York Art world' circa 1985. The real-world effects of this campaign are recited like a charm or a spell that does not so much evoke the past in the present as preserve it into the future of feminist art activism:

> Who knew that our work would cause all hell to break loose? Who knew it
> would cause a major crisis of conscience about diversity in the art world … Who

DOI: 10.1057/9781137374851.0007

knew that those two posters would lead to hundreds of others... Who knew that 25 years later we – the agitating outsiders – would wind up inside the museums we criticize... and be speaking to all of you today?[4]

The Guerilla Girls' incantation of their unpredictable impact switches to a series of exhortations addressed to 'the class of 2010', conceived of as creators and strategists in art, design, writing, administration, architecture and education. Both listed and narrativized, these exhortations include:

> **Be crazy**. Political art that just points to something and says 'this is bad' is like preaching to the choir. Try to change people's minds about issues. Do it in an outrageous, unforgettable way. A lot of people in the art and film world didn't believe things were as bad as we said they were and we brought them around... with facts, humour and a little fake fur.

'Facts, humour and a little fake fur' is a phrase that nicely encapsulates the tension between rightness and writerliness in contemporary manifestos that draw on an expanded notion of writing (writing beyond literal inscription) while being predominantly written (literally inscribed).

Tkacova and Chisa's manifesto takes the form of a short video showing a troup of marching majorettes performing an apparently generic choreographed routine. However, as Clara Orban notes[5], their movements are actually semaphore, coded language that spells out or rewrites the concluding part of 'The Manifesto for Futurist Woman', Valentine de Saint-Point's 1912 response to F. T. Marinetti's 'The Futurist Manifesto' of 1909. Tkacova and Chisa's work of homage is followed up in 2011 with a piece entitled '80:20', another critique of the art institution's asymmetry and lack of diversity developed in this case for the Venice Biennale. Presented in list form on the façade of the Romanian Pavilion, the project offers an 80:20 position on the reasons to be, or not to be at the Venice Biennale. Despite being literally inscribed in stone, the project recognizes the slipperiness of the list, or the fact that 'the reasons "against" can be seen as the reasons "pro" and vice versa'.[6] Martine Syms writes for and against science fiction, countering the spectacular with the vernacular, the 'stupidities' of Eurocentric fantasy ('Jive-talking aliens'; 'Inexplicable skill in the martial arts') with the quest for black diasporic artistic production and re-imagined futures. Drawing on a sense of contingency and incongruity, Syms seeks a diverse, polyphonic Mundane Afrofuturist literature devised according to a set of negative rules:

DOI: 10.1057/9781137374851.0007

1   No interstellar travel – travel is limited to within the solar system and is difficult, time-consuming and expensive
2   No inexplicable end to racism – dismantling white supremacy would be complex, violent, and have global impact
3   No aliens unless the connection is distant, difficult, tenuous and expensive – and they have no interstellar travel either
4   No internment camps for blacks, aliens, or black aliens
5   No Martians, Venusians, etc.
6   No forgetting about political, racial, social, economic, and geographical struggles
7   No alternative universes
8   No revisionist history
9   No magic or supernatural elements
10 No Toms, Coons, Mulattoes, or Bucks
11 No time travel or teleportation
12 No Mammies, Jezebels, or Sapphires
13 Not to let Mundane Afrofuturism cramp their style, as if it could
14 To burn this manifesto as soon as it gets boring[7]

I accept these rules as my own, and undertake to follow them in the continued pursuit of my own, 'mundane' science fiction, my own ordinary, every day in the life of Janet Smart. Syms' rules are applied to her own story, entitled 'Most Days'. This explores an average day in the life of a young black woman in a Los Angeles of the future. Written as a screenplay, it has been recorded along with an original score by her collaborator, Neal Reinalda.[8] Syms' screenplay, like my novel in progress, is co-extensive with the manifesto form, a form constituted by writing as both a specific and non-identical technology whose role, or strategic intent is de- and reconstructive, including of writing itself.[9]

To what extent does the inherent writerliness of the manifesto contribute to its continual reforming and remediation, extending it through videos, screenplays and novels and rendering it, perhaps, self-reflexive or even diffracting[10] what the manifesto – what writing the manifesto – will do? Without this perspective on writing, it would be easy to exaggerate the difference between early twentieth-century, and early twenty-first century manifestos and to suggest, for example, that they changed from inciting violent, revolutionary opposition to the status quo to advocating and embodying more subtle forms of struggle. The language and tone of, for example, Valentine de Saint-Point's 'The Manifesto of Futurist Woman (Response to F. T. Marinetti)' (1912) and Mina Loy's 'Feminist Manifesto'

DOI: 10.1057/9781137374851.0007

(1914) is certainly radical, rooted in a dialectics of sex and gender, nature and culture, body and machine, past and future that is subsequently broken down. However, it is also knowingly hyperbolic, poetic and, in de Saint-Point's case, parasitic with regard to Marinetti, manifestos and men in general. Hers is a 'contrary exaggeration'[11] to his 'contempt for woman',[12] one that takes it on – 'Enough of those women, the octopuses of the hearth, whose tentacles exhaust men's blood and make children anemic' – in order to redirect it – 'Women are Furies, Amazons... combative women who fight more ferociously than males...' Rather than positing a then and now of the manifesto, transformed from an attitude of violence to one of humour, polemic to playfulness or even anti- to pro-feminism,[13] I would argue that it continues to evolve into, and as an antagonistic form by means of writing about and writing out of social and historical contexts that remain structured by gendered hierarchical dualisms.

Parody and irony are particular writing strategies – characteristic of, for example, Valerie Solanas' 'SCUM Manifesto' (1967) and Donna Haraway's 'Manifesto for Cyborgs' (1985) – that serve as hinge points between description and reinvention, art and activism, critique and creativity, writing about and writing out. Arguing that Solanas parodies the manifesto form itself while Haraway's manifesto is more ironic, Janet Lyon implies that parody operates as a more effective hinge point than irony (1999). Parodying 'in outrageous caricature, the formal aspects of the political manifesto', Solanas exposes the contradictions between 'violent threat/static word and rationality/rage', and offers 'a recipe for more literal violence and less orderly rage, served up by "females" who claim it as their political right' (173). Her address to these 'thrill-seeking females'[14] – 'The "we" of SCUM is Valerie Solanas and People Like Her; but she is the sole signatory of the tract' (Lyon 1999: 172) – is, Lyon argues, more direct and effective than Haraway's ironic meditation on political identity (the heart and central problematic of the manifesto) embodied in the non-unitary figure of the cyborg:

> In calling for a complicated ironic myth rather than calling an audience to action, 'Cyborgs' in part highlights its wariness about manifestic discourse: it cautions against feminist taxonomies that 'tend to remake feminist history to appear to be an ideological struggle among coherent types persisting over time... [with the result that] all other feminisms are either incorporated or marginalized'.

> (Lyon 1999: 195)

DOI: 10.1057/9781137374851.0007

What the cyborg's liminal, dynamic subject-object positioning does is complicate the homogeneous, singular or collective pronoun, the I/we of the manifesto, opening it to 'multiple political affinities (not identifications)' and coalitions 'ultimately with the aim of forging a nontotalizing, degendered, anticapitalist global politics' (195). Insofar as this politics is envisioned through coalitions and affinities, it also, I would argue, allows for conflict and tension to exist precisely in the cyborg's ironic reference to a figure Haraway identifies with the military industrial complex. In her political myth or imaginary, there are cyborgs and then there are cyborgs. Evoking the cyborg is ironic in the sense that:

> Irony is about contradictions that do not resolve into larger wholes, even dialectically, about the tension of holding incompatible things together because both or all are necessary and true. Irony is about humour and serious play. It is also a rhetorical strategy and a political method, one I would like to see more honored within socialist-feminism.

> (Haraway 1991: 149)

Noting, as Haraway subsequently does[15], the context-dependence and therefore redundancy of the cyborg figure (as a figure of Cold War, East-West conflict), what more can be done to honor the cyborg's principal technology of writing as a paratextual strategy, a mechanism for re-wording and re-worlding and for re-telling stories that matter? Insofar as they ever did, do irony and parody still work? Irony as non-dialectical opposition becomes less evident in Haraway's successor figurations – the modest witness is an ironic figure of detached observation and disembodied knowledge, but the companion species and in particular the dog is much less of an antagonist. However, irony might still work for my re-gendered *i*media subject, always already at risk of non-identity and forged in contradictory contexts of un/sustainability, of both potential and *potentia*. I argued in the previous chapter of this book that potential and *potentia* signal and evoke conflicting notions of time, namely, the increasingly finely grained and ingrained clock time that carves out women's work in multiple, highly differentiated geographical, economic and social realms and the life-times of women's diverse becomings. The subject of my own manifestos is, like Haraway's, the subject of my manifestos. She is what I'm writing about, as well as being-becoming a contingent, provisional figure through which I seek to write out, to change or transform the contradictory states 'we' are in and to imagine a politics, specifically a feminist politics that is post-dialectical (not in

DOI: 10.1057/9781137374851.0007

the sense of being after) and therefore less susceptible to the totalizing force of neoliberalism. If the dialecticism I've outlined in earlier chapters is compatible[16] with what Mouffe describes as a shift from revolutionary political discourse to an 'exodus' from political discourse, for example in favor of objects, environments and materials as (ultimately marketable) things-in-themselves, then the quest for a post-dialectical feminism places political subjects in tension with marketable objects in a way that remains dynamic, irresolvable and characterized less by opposing id/entities than by the possibilities of movement itself. Rooted in deconstruction, post-dialectical feminism is not about the substitution of subjects for objects as much as it is about the possibility of moving beyond this and other dichotomies. The quest for a post-dialectical feminism, which I will outline in more detail here, is inherent in feminist deconstructionism and traceable between Haraway's cyborg manifesto and Cixous' 'The Laugh of the Medusa'. By arguing that Haraway falls short of the manifesto proper, confining herself to 'a theoretical preamble to the manifestic speech act', Lyon (1999: 195) dishonors some of the founding work of feminism, failing 'to admit that writing is precisely working (in) the in-between' (Cixous 1985: 254) and underestimating the task of re-writing the manifesto's political subject so that, as Cixous herself recognizes, she can move:

> As a militant, she is an integral part of all liberations. She must be farsighted, not limited to a blow-by-blow interaction. She foresees that her liberation will do more than modify power relations or toss the ball over to the other camp; she will bring about a mutation in human relations, in thought, in all praxis.

> (Cixous 1985: 253)

Writing as queer feminist praxis   DEFINITIONS

To argue that writing is a queer feminist praxis is in part to challenge a certain 'citation practice', one that is associated with a particular form of dialectical feminism related to, but distinct from, the division between epistemology and ontology (or between subjects and objects of knowledge) (Hemmings 2014: 29). This is the so-called reparative turn against critique as a hermeneutics of suspicion or a kind of paranoia (2014). Eve Kosofsky Sedgwick argues for a turn from a paranoid disengagement from the world and its objects that has done nothing, achieved nothing except the false elevation of a knower who is above the beyond her

DOI: 10.1057/9781137374851.0007

objects (untouched by them) to a reparative relation to, or with them (Wiegman 2014). The structure of such an argument seems to me not only to reproduce the very problem of paranoia as splitting – and indeed projection – that it claims to distance, but to also miss what, for Melanie Klein, whose psychoanalytic theory is evoked here, is the point of reparation, namely to arrive at a developmental stage[17] of what Haraway would call irony, a holding together, in permanent tension of feelings of love and hate, suspicion and trust, judgment of and engagement with the (original) object. The reparative turn might turn away from writing's paranoid, interpretive past, or alternatively claim it for a future of affective engagement but, as Clare Hemmings suggests, writing in fact signals a way out of a false dichotomy that is aligned with a split between feminist epistemology and queer theory respectively (2014). Hemmings finds herself:

> Querying the implicit – or explicit – positioning of queerness on the side of everything that is transformative or creative, while poor old feminism is consigned to the reproduction of dull – and already fully known critique. In the sexual division of theoretical labour, queer theorists and not feminist theorists appear to be having all the fun. And further, in such iterations of the history of theory, feminism frequently emerges as the dominant force in need of displacement if we are to have a creative future.
>
> (Hemmings 2014: 29)

If writing remains undecideable with respect to the divisions that Hemmings exposes[18] and if, more importantly, it works to dismantle them through an inventiveness that operates between interpretation and affect[19] then the question, as Cixous recognized, following Derrida, is not what writing is but what it does under specific conditions of possibility (Kember 2014). It will always be impossible, Cixous argues, to define and theorize what writing – in her case, feminine writing – is, but what it does, under conditions of phallogocentrism is sow disorder and bring about upheaval. There is, she says, 'no other way'. There is no room for feminine writing within a phallogocentric order, 'no room for her if she's not a he'. But if 'she's a her-she, it's in order to smash everything, to shatter the framework of institutions, to blow up the law, to break up the "truth" with laughter' (1985: 258). I don't think Haraway's approach to writing, albeit one that appears to be much less destructive, contradicts Cixous' as much as it builds on it as the principal mechanism of deconstruction that corrupts order – in the form of binaries – in order to bring something

DOI: 10.1057/9781137374851.0007

else about, to introduce new, provisional concepts – such as natureculture, FemaleMan and OncoMouse – without ever seeing them as a way
of resolving a dichotomy (Haraway 1991; 1997). Haraway's re-wording
strategies are re-worlding strategies in process, in action, in the present
participle (Haraway in Gane 2006; Weber 2000). They do not resolve
conflict but create openings for political alternatives by moving between
– toward and away – from the rock and the hard place that Cixous sees as
Medusa and the abyss: 'Let's get out of here' (1985: 255). Ultimately, writing as *the* deconstructive mechanism is more of an ongoing, unfinishable
movement (a process of diffraction rather than a diffracted con/text)
than a method or mode. Its 'strategic intent'[20] is delimited by the structures on which its work depends, leading to the familiar and frustrated
critique of critique (of the critical tradition), namely, that it fails to create
alternatives, change, or really anything much at all. Haraway succumbs to
this critique by creating what, for me, would be a false division between
deconstruction and dreamworlds (Gane 2006), one which, in effect,
continues to structure aspects of feminist, queer, cultural and *i*media
theory and in doing so, elides conflict as the central problem of political
theory, substituting it, as it were, by the substitution of this for that.

Key aspects of queer feminist writing praxis include parody and
irony. While their definitions and differences have long been explored
and debated (Jameson 1991; Hutcheon 1989 and 1994; Colebrook 2004;
Dentith 2000), their efficacy arguably remains contingent on conflict,
on the underlying but irresolvable tension that Jameson saw as critical
distance from the ur-text (1991). Moving uncertainly between subversion
and consent, what Colebrook (following Hutcheon) refers to as ' "the
communicative space" of irony' – as well, I would suggest, of parody –
remains unstable as the conditions of possibility for this kind of writing
continue to evolve (2004: 158). The fact that in the UK there are now
copyright exceptions for parody[21] (particularly online forms) is not a
stable indication of conservatism or of creativity in cultural and political contexts in which such impulses are thoroughly enmeshed, notably
through the dominance of concepts like innovation. Indeed, just as such
resolutions are offered within neoliberalism, the work of parody and
irony, despite or rather because of their mobility and undecideability, may
be highlighted as a way of recreating tensions – more or less. For me, this
cultural-political work remains literally vital or both alive and expedient
even as it goes wrong, the tension becomes less and less and antagonisms
morph into the sort of moralism that produces non-ironic, ironic sexism

DOI: 10.1057/9781137374851.0007

and his stories of misogyny. I read these *hi*stories, with Laura Bates and Anita Sarkeesian in games and streets and in schools and workplaces as well as in industry-led visions of 'our' EuroAmerican pale and male-centric so-called smart futures. I also read them, and seek to write out of them in *hi*stories 'proper', in novels such as Gary Shteyngart's *Super Sad True Love Story* (2011). This witty, dystopian account of mishearing voice recognition technology and biotech fantasies of Indefinite Life Extension[22] set in a future U.S. of parodically heightened paranoia and plummeting financial authority centers on the relationship between an apparently self-deprecating, middle-ageing male of the old-world (he still likes books) and a cold, money-oriented, too-young woman who thinks his books smell and who seems to embody the author's ideas of all that is wrong with the configuration of technology, control and consumerism. As she, Eunice Park, emotionally and financially destroys the protagonist, Lenny Abramov, Shteyngart drives the plot through national and global apocalypse and personal and political tragedy that was always already all about 'him'. As the cultured, bookish, ironically (in the sense of not meaning what he says) self-hating Lenny lives out his three-point plan to seduce Eunice, using a Chekhov novella as his guide,[23] the shallowness of Eunice's world – all social networking, personal rating, credit ranking and clothes buying – is laid bare, just as she is, first through her obsession with revealing underwear and then through the peeling away of her onionskin jeans with their 'rigid, empty skins' (207). Eunice, and/ as the world that Lenny non-ironically hates, first mediates and then unmediates it for him, rendering graphically clear and transparent – as if she and it were made of glass – a promise of openness and availability already foreclosed. He sees through Eunice as she shops:

> Her face was steely, concentrated, the mouth slightly open. Here was the anxiety of choice, the pain of living without history, the pain of some higher need. I felt humbled by this world, awed by its religiosity, the attempt to extract meaning from an artifact that contained mostly thread. If only beauty could explain the world away. If only a nippleless bra could make it all work.

> (Shteyngart 2011: 207)

Then, through the telling of his story – Lenny publishes his diaries and Eunice's text messages – he survives the apocalypse and prevails. I find it interesting that Shteyngart's character precludes the retelling of his story – one in which Eunice's writing is seen by the critics as a 'welcome relief from Lenny's relentless navel-gazing' (325) – by telling the actors in a

DOI: 10.1057/9781137374851.0007

possible video adaptation ('they had no idea who I was') that the central protagonists and short-term antagonists were dead: 'I laid out a scenario for the final days of Lenny Abramov and Eunice Park more gruesome than any of the grisly infernos splashed on the walls of the nearby cathedral' (329). So much more promising than the civilized but judgmental tone of the novel as a whole, the violence of Lenny's last, incongruous words nevertheless augur a retelling of the tale, one that is more about her than him. After all, Eunice is not dead. She is living with someone she met at Goldsmiths. Moreover, she is, by critical consent, the better writer of the two. According to one review:

> 'She is not a born writer, as befits a generation reared on Images and Retail, but her writing is more interesting and more alive than anything else I have read from that illiterate period ... what comes through is a real interest in the world around her – an attempt to negotiate her way through the precarious legacy of her family and to form her own opinions about love and physical attraction and commerce and friendship ...'
>
> (Shteyngart 2011: 325–6)

Toward a post-dialectical feminism   DIALECTICS

I am interested in an ongoing movement towards a post-dialectical feminism that engages its boundary work with political theory and in doing so, directs outwards (through anger and laughter) its 'anxiety about loss of identity and place' (Brown 2005: 65). The quest for a post-dialectical feminism must recognize its own futility as well as necessity. Here, post could never mean after. It is impossible, as Derrida states, to oppose or to substitute dialectics for something else:

> Be it opposition to the dialectic or war against the dialectic, it's a losing battle. What it really comes down to is thinking a dialecticity of dialectics that is itself fundamentally not dialectical.
>
> (Derrida 2001: 33)

Thinking a dialecticity of feminist dialectics that is itself fundamentally not dialectical might entail an attention to the movements between, or toward and away from, poles. It might entail writing about and out of socio-historical contexts that incorporate and produce feminisms themselves and the undoing of what has been undone; the reversal

DOI: 10.1057/9781137374851.0007

and displacement of post-politics – or 'serious political nihilism' – and non-ironic, ironic post-feminisms such as Shteyngart's (Brown 2005: 59). Thinking in this way is about the retention of a tension that Mouffe terms antagonism and that 'impedes the full totalization of society' just as it 'forecloses the possibility of a society beyond division and power' (2013: 1). If Mouffe's concept of agonism or antagonistics seeks to convert violent conflict into the plural counter-hegemonies of a radical democracy, envisioning relations between adversaries rather than enemies, it does not dispense with conflict altogether: 'In an agonistic politics... the antagonistic dimension is always present, since what is at stake is the struggle between opposing hegemonic projects which can never be reconciled rationally, one of them needing to be defeated' (9). An agonistic politics, rooted in conflict and dissent, retains an adversarial relation, rather than one based on substitution, with rival political projects and theories including those that are rationalist or 'deliberative' and individualistic (6). What does this mean for feminist politics and for feminist political theory as it continues to evolve along paths more genealogical than teleological? Is it necessary to trace the non-linear history of theories based on agonism versus associationism,[24] passionate dissent versus rational consent, negativity versus affirmation, etc.,[25] or can feminism come to terms with its own adversarial relations, predicated perhaps on the 'conflictual consensus' of constitutive outsides in order for its extended subject-object to critically engage, in more mature forms, with 'the world around her'? (Mouffe 2013: 8; Shteyngart 2011: 325). If the developmentalism of Klein's psychoanalytic theory were bracketed,[26] could feminism's political subject be (re)conceived within the intractably difficult, ambivalent 'phase' that Klein refers to as 'depressive anxiety' (1988: 36)? For Klein, this phase or position refers to border states, including those between internal-psychic and external-social worlds (20). It incorporates guilt (toward the previously persecuted and persecutory object), grief and mourning but also growth, integration and understanding. Taking the border state as a performative metaphor for feminism itself (for feminisms themselves), the object of loss, grief and mourning may be the idea(l) future of a 'final ground' for feminist politics while the outcome of growth, integration and understanding might be the sense of feminisms as continual openings to the political (Mouffe 2013: 2).[27] The distinction Mouffe draws between politics and the political is clear but not absolute: 'By "the political", I refer to the ontological dimension of antagonism,' and by "politics" I mean the

DOI: 10.1057/9781137374851.0007

ensemble of practices and institutions whose aim is to organize human coexistence. These practices, however, always operate within a terrain of conflictuality informed by "the political" ' (2013: xii).

Conclusion

Feminist political theory-practice ⎡DECISIONS, DECISIONS⎤

What is at stake in reading-writing feminism as both politics and the political in a way, perhaps, that Brown is more inclined to do than Mouffe herself?[28] Brown's sense of 'feminism unbound' inquires into possible openings into the political after a politics predicated on sex and gender. Importantly, not least for my argument about the re-gendering of *i*media, being 'after' sex and gender here does not mean ' dispensing with them but rather, perhaps, is more like being after the Fall, after their Fall. Fallen yes, but like all toppled sovereigns and overthrown founders, they do not thereby cease to govern' (2005: 98). For Brown, the question is not whether feminism can survive without sex or gender 'but *how it lives*, and will continue to live, without a revolutionary horizon' (99). Relevant to this question of how feminism lives is Mouffe's characterization of the trajectory taken by radical projects that, in her view, forsake 'the traditional revolutionary approach' for another one that 'under the name of "exodus", reproduces, albeit in a different way, many of its shortcomings' (2013: xiii). Mouffe contests:

> The total rejection of representative democracy by those who, instead of aiming at a transformation of the state through an agonistic hegemonic struggle, advocate a strategy of deserting political institutions. Their belief, in the availability of an 'absolute democracy' where the multitude would be able to self-organize without any need of the state or political institutions signifies a lack of understanding of what I designate as 'the political.'

> (Mouffe 2013: xiii)

Mouffe refers specifically to the beliefs of Michael Hardt and Antonio Negri, whose argument, for example, in their book *Empire* is based on a philosophy of immanence unable, she argues, 'to give an account of radical negativity, i.e., antagonism' (78). While she does not elaborate this underlying philosophy or ontology of immanence in her work on agonistics, Mouffe's sense of thinking the world politically is contingent on a departure from it and from the relational ethics that is – albeit

DOI: 10.1057/9781137374851.0007

elsewhere[29] – associated with it. The 'current zeitgeist' may favor such an ethics, but it is one that tends to valorize multiplicity and relationality as if power were not in play. But 'when we acknowledge that antagonism is ineradicable and that every order is an hegemonic order, we cannot avoid facing the core questions of politics: What are the limits of antagonism, and what are the institutions and the forms of power that need to be established in order to allow for a process of radicalizing democracy?' We cannot, Mouffe suggests, avoid the 'moment of decision' that will always involve some kind of closure or exclusion (15). A politics based on the disturbance or destruction of existing orders is not enough, since decisions and closures, or what in ethical parlance we might term 'cuts' still need to be made. In *Life After New Media*, my co-author and I theorized such cuts in the space between ethics and politics, bringing together the philosophy of immanence with that of transcendence and confronting vitalism with the critical tradition in order, precisely, to do more than disrupt traditional approaches to media and to explore the possibilities of thinking and doing media differently (Kember and Zylinska 2012). I am wary of some of Mouffe's own distinctions or oppositions – including that between ethics and politics – that might stifle rather than nurture the liveliness of objects of thought and action such as media and feminism. In thinking these objects together in this book, I have emphasized one particular mode, method or rather movement of cutting, namely, writing.

Feminist writing, as both a continuation and transformation of feminine writing is both destructive and co-creative of *i*media worlds. Its (though there is of course no 'its' here, no singular notion of feminist writing notwithstanding shared strategies of reinvention) extended subject-object is post-human only in the sense of not being after the human and, as well as being re-differentiated is also conflicted, contingent, in my account, on the contradictory time zones of potential and *potentia*, productivity and becoming. She is the tension between *homo oeconomicus* and its constitutive outside. At once singular and multiple, she has a given name and narrative that through writing might be reclaimed and retold. The manifesto is one among other old and ever new provisional forms of writing that combine and recombine rightness and writerliness, generating ironic and parodic openings to the political – more or less. Such openings are themselves undecideable, never fully or finally open or closed. In as far as they foreclose on the fantasy of a final ground (a future, alternative, perhaps affirmative feminist politics)

DOI: 10.1057/9781137374851.0007

they do so in favor of a non-totalizable, post-revolutionary feminism constituted by border states – internal and external, psychic and social – and by writing that is about and out of the socio-historical contexts that are its conditions of possibility. Writing, as an affective-interpretive queer feminist praxis was always about letting our objects fly (Brown 2005: 115). It was always about and out – imaginatively, transformatively – of identity, of sex and gender. If, with Cixous, we continue to speak about writing, 'about what it will do,' we exclude as unnecessary the question that Mouffe considers supplementary to that of identity and its proliferation. This question 'of what we should do as citizens' has an answer: 'write' (Mouffe 2013: 12; Cixous 1985: 247).

Notes

1   This question has been asked before, by Hélène Cixous in 'The Laugh of the Medusa', *Signs*, Summer 1976 ('I shall speak about women's writing: about *what it will do*'). See also Cixous, H. (1985) 'The Laugh of the Medusa', in E. Marks and I. de Courtivron (eds) *New French Feminisms,* Sussex: The Harvester Press Limited.

2   My emphasis so far has been on the association between rightness and masculinism or phallogocentrism – what Xavière Gauthier refers to, in 'Why Witches?' as 'The frightful masculine fashion of ... speaking in order to be right – how ridiculous!' (in E. Marks and I. de Courtivron (eds) 1985: 200).

3   This list was selected almost entirely at random and is based on the final entries compiled by KT Press for n.paradoxa international feminist art journal www.ktpress.co.uk/feminist-art-manifestos.asp

4   www.guerillagirls.com

5   http://www.chitka.info/manifest.html

6   www.chitka.info/8020html

7   http://rhizome.org/editorial/2013/dec/17/mundane-afrofuturist-manifesto/

8   Mixed Media Recordings, 2014.

9   Writing of course is also the principal mechanism of deconstruction: 'The process of writing always reveals that which has been suppressed, covers over that which has been disclosed, and more generally breaches the very oppositions that are thought to sustain it' (Reynolds, J. (2010) 'Jacques Derrida (1930–2004)', Internet Encyclopedia of Philosophy http://www.iep.utm.edu/derrida/)

10   Haraway insists on the difference between self-reflection and diffraction in a passage from her book *Modest Witness* that is worth citing at length. The

DOI: 10.1057/9781137374851.0007

difference has, in short, to do with categories of thought that do or do not make a difference:

> My invented category of semantics, *diffractions*, takes advantage of the optical metaphors and instruments that are so common in Western philosophy and science. Reflexivity has been much recommended as a critical practice, but my suspicion is that reflexivity, like reflection, only displaces the same elsewhere, setting up the worries about copy and original and the search for the authentic and really real. Reflexivity is a bad trope for escaping the false choice between realism and relativism in thinking about strong objectivity and situated knowledges in technoscientific knowledge. What we need is to make a difference in material-semiotic apparatuses, to diffract the rays of technoscience so that we get more promising interference patterns on the recording films of our lives and bodies.
>
> (Haraway, Donna J. 1997: 16)

11   http://www.wired.com/2008/11/the-manifesto-1/

12   http://vserver1.cscs.lsa.umich.edu/~crshalizi/T4PM/futurist-manifesto.html

13   de Saint-Point's statement that 'feminism is a political error' remains contestable within the political theory of, for example, Wendy Brown and Chantal Mouffe. Any supposed transition from anti- to pro-feminism would fail to take account of the extent to which manifestos negotiate feminist politics within complex, overlapping as well as changing social and historical contexts.

14   http://www.womynkind.org/scum.htm

15   See for example, Gane, N. (2006) 'When We Have Never Been Human, What Is to Be Done? Interview with Donna Haraway', *Theory, Culture & Society*, 23 (7–8): 135–8.

16   I have deliberately chosen to refer to compatibility rather than complicity in order to avoid being accusatory, judgmental or moralistic and to remain open to the prospect of antagonism which, for Mouffe, impedes the full totalisation of society (2013).

17   I would say, not only psychologically, but also culturally and even theoretically.

18   Hemmings is responding to Robyn Wiegman's argument in 'The times we're in: Queer feminist criticism and the reparative 'turn'', *Feminist Theory*, Volume 15, Number 1, April 2014: 4–27.

19   Interpretation as such is impossible within a deconstructive reading-writing strategy. Encounters with texts and contexts are neither faithful nor transgressive but interventional and inevitably inventive.

20   Reynolds, J. (2010) 'Jacques Derrida (1930–2004)', Internet Encyclopedia of Philosophy http://www.iep.utm.edu/derrida

DOI: 10.1057/9781137374851.0007

21 Research shows that parody channels revenue back to the source and does not significantly damage the economic interests or reputation attached to the ur-text (Erickson, K. Kretschmer, M. and Mendis, D. 2012).

22 Or, 'effeminate life invention' to the ears of a not altogether benign otter avatar (7).

23 *Three Years* is 'the story of the unattractive but decent Laptev, the son of a wealthy Moscow merchant, who is in love with the beautiful and much younger Julia' (34).

24 'Associational theorists... gravitate more toward deliberative democratic theory, while agonistic theorists... worry that democratic theories that focus on consensus can silence debate and thus they focus more on plurality, dissensus, and the ceaseless contestation within politics' (www.plato.stanford.edu)

25 See Butler on 'etc.'

26 Although it should be noted that Klein's sense of the succession of paranoid/schizoid and depressive positions is non-linear (Klein 1988).

27 This is, or these are the feminisms to which Haraway refers, citing bell hooks: 'In bell hooks' terms, as a verb, feminism is about women's moving, not about some kind of particular dogma' (in Gane 2006: 136).

28 Brown approaches feminism in its constitutive relation to political theory where Mouffe appears to regard the interests of feminism – of agonistic feminism – as being too limited to questions of identity (2013).

29 See for example, Barad (2007) and Braidotti (2006). Also, Kember and Zylinska (2012) combine immanentist and transcendentalist traditions in order to examine the political and ethical implications of human relations with media.

DOI: 10.1057/9781137374851.0007

6

interlude 2: Excerpt from *A Day in the Life of Janet Smart*

Abstract: A Day in the Life of Janet Smart *is a novel in progress that temporarily occupies Philip K. Dick's* Minority Report *and parodies the extent to which its predictive-responsive i-enabled glassworlds inform corporate and computational visions of our smart futures. Skeptical about what these futures have already delivered in terms of intelligent labor-saving devices like care robots and self-driving cars, Janet starts her day with a serious case of jet lag, juggles her job, her professional, personal, human, non-human and casual relationships and then has to figure out how to stop her daughter living the same unsustainable life on a whole other unsustaining planet.*

Kember, Sarah. *iMedia: The Gendering of Objects, Environments and Smart Materials*. Basingstoke: Palgrave Macmillan, 2016. DOI: 10.1057/9781137374851.0008.

DOI: 10.1057/9781137374851.0008

Janet in the frame

She must have nodded off. She was still sitting upright with her eyes open and her hands perched over the embedded keyboard. Her mother had always insisted on good posture, and when she was growing up, a rather severe piano teacher who would not countenance slouching of any pertinent body part had reinforced this lesson. The review screen brightened in recognition of her eye movement and, being networked with Colin, displayed the time just as he had, in all of her preferred locations. Only ten minutes and twenty seconds had passed since Marvin had gone on his break, but she'd been dreaming about Jennifer, having one of those dreams that cannot be described as a nightmare but that actually upset you, erasing the distinction between sleep and wakefulness. They'd been in Jennifer's bedroom, packing a suitcase for her space flight. Jennifer was still seven years old, but her walls had grown up and sprouted fantastic blooms in purple and red. Zachary was over in one corner, stationary and defunct. The suitcase was old, inherited from Janet's grandfather who had kept his toy cars in it. It was also small and so they were trying to decide what to take and what had to be left behind. Janet wanted to climb in it but it was already half full of carefully folded clothes. They were adult clothes, and Jennifer was talking excitedly about the microbiological samples they'd be transporting and that it would be her job to preserve. Janet had tried to make encouraging noises but some other sound escaped her, something primal that shocked them both and led to an argument that quieted now as she faded in to consciousness with one clear thought: how to prevent her daughter from embarking on a journey she could never return from.

Janet had known about the Mars mission before today but had given it little thought. To her, it was a longstanding, somewhat tired human conquest story that belonged in books and films and was occasionally in the news. Who knew if that meant it was any more likely to happen? The habitation of Mars is something that had always – and never – been happening. It was a kind of permanently delayed but historically familiar tale of exploration and the colonization of worlds that had never gone particularly well for the locals. As far as she knew, scientists had given up looking for actual Martians years ago and now they were planning to import them. It had never sounded like a good idea, even before it became personal. If Mars wasn't fit for the sort of microbe that had been around for billions of years, it was hardly going to be fit for humans who

DOI: 10.1057/9781137374851.0008

were much younger, in evolutionary terms, and needed rather more in terms of sustenance. This was going to be an experiment in survival against all odds and any common sense. It was for dreamers like Marvin who couldn't or wouldn't see that they were being used, experimented on. There would be a social side to this too, the settlers inevitably becoming stars of some hideous virtual-reality show. Janet would have to watch as the mission and its entire community, including her own child, unfolded and unraveled in real-time. If they took a PreCon... but no, why would they? Jennifer's sacrifice would be made in the name of science, not commerce, and in either case it was the sacrifice bit she was having trouble with. Perhaps the solution was for her to resign from the agency, stay at home and teach her daughter about anything and everything that might divert her interest. If it was too late for that, then she should stay put and find a way to use her influence. She might not have the power to scupper an IASA mission, but she could do something to expose the lie about the technology it depended on.

Janet was managing to make herself feel better and worse at the same time. She felt a bit more resolute and also guiltier about Jennifer, whose dreams would not be negated so much as curtailed, offered a boundary, a limit in place of a frontier. She spoke her name three times before the correct face appeared on her screen. The janitor had looked confused when she called him darling and her old friend Jane seemed delighted to see her when she was prematurely cut off. I'll call her back when I get a minute, Janet told herself. Jennifer was in fact looking a lot better. Suspiciously so, Janet thought. It was almost as if she'd figured out a way to fool everyone. Not that it would take that much, since everyone, for Jennifer, consisted of an ageing machine (albeit one with enough cunning to keep her test results to itself) and an absent mother. Janet knew she had no right to be angry about this minor deception but she couldn't help being piqued by the apparent conspiracy and the fact that Jennifer enjoyed hanging out with Zachary more than going to school. She attributed her daughter's inability to get along with other children to external factors, to nurture rather than nature. This meant that she would simply have to be more vigilant in monitoring her attendance as well as her extracurricular activities. With a little less input from Zachary, it was perhaps possible, after all, to turn things around. Janet was not even averse to asking him to teach Jennifer how to paint, draw or better still, play the violin. She pictured the scene of aural torture and a little snort broke the silence in the room.

DOI: 10.1057/9781137374851.0008

As Senior Operator, Marvin's job was to rouse Polly from her naps and review whatever data they had enabled her to process. The naps were a kind of unclogging mechanism, designed to keep information flowing but without any concomitant effect on mood. The latest PreCons had better-integrated circuits and required fewer naps. Marvin often reflected on this when he went to wake Polly and she reacted as if he was trying to murder her. For the most part, he was rewarded with what he considered to be low-grade material or what the agency referred to as minor reports. Anything of real interest, a major report, would override Polly's dormant phase immediately. Marvin had programmed this override himself because Polly's emotional states were fixed at intermediate, and so she could respond with equanimity-excitement but not one or the other. An alarm – and not always a false one – would go off on his watch, and he would be plugged in to what Polly was seeing. It had happened, for real, once or twice when the unmanned aircraft went commercial and Moogle Earthtechnics was formed from the merger of the two remaining independent technology companies. Marvin was wearing his MEwatch6 now, and he checked it, tapping the screen and adjusting the volume to see if somehow he'd missed something. Polly wasn't only awake; she was standing there staring at him.

'Janet.'

'Marvin.'

'You need to see this.'

Janet had been looking at her late morning schedule. The first item was classified as a minor report, but there was nothing after it – nothing at all. The classifications, she knew, were post facto and courtesy of the sorting algorithm Marvin had written. Out of frustration at being unable to filter the data that came through Polly, he had sought to organize what came out. The balance was certainly not what he wanted, but the sorting had never gone wrong before. Something always came out. Their caseload was decreasing, but it would not have suddenly stopped. Polly was stuck.

'She did her "who are you, don't hurt me" routine and then she froze. Nothing,' Marvin said, quietly, as if he was afraid to disturb her, wherever she was.

Janet suppressed the thought that she might not be able to pop home at lunchtime as she'd hoped and approached Polly. If Marvin found Polly's emotionality excessive, grating if not exactly detrimental to her rational functioning, Janet considered it to be sometimes comically, sometimes painfully limited. Marvin's reactions left little room for her own, and so she accommodated them, recognizing that although Polly was at one with all of the devices

DOI: 10.1057/9781137374851.0008

that bigged up their part, she, if not she alone, had a part, and the agency would not exist without her. Janet respected her functionality but had taken for granted how even a formulaic and constrained emotional intelligence had raised her above this and created something seemingly animate. Without speech, movement or expression, Polly was divested, revealed in her non-being. There was no point staring into her eyes that were lenses or shaking her mechanical frame. Janet, it must be said, was not particularly technically minded. Still, she knew the one thing that everyone knew. Polly's brain was a computer after all.

'Switch her off Marvin. We'll reboot her.'

Marvin pulled a Polly face. 'You are kidding me? That is a joke?'

'No.'

'She's an artificial life form, Janet ...'

'... which at this moment is a malfunctioning machine.'

'Wrong. Machines can be turned on and off. Alife's are different, sentient.'

'She looks it.'

'It can't be done, and it wouldn't be right.'

'There is no right and wrong here, no morality to apply. We need to restore her function, and you don't seem to have a better idea.'

'I do. We wait. She'll come round.'

They both looked at Polly for a couple of minutes and then Janet turned to Marvin. 'Well?'

'She must still be processing, like an impromptu nap. Worst case scenario is some kind of shock-induced coma.'

'Coma? So what, we call someone with a white coat and a stethoscope. That's your idea?'

'No of course it isn't, but you don't just pull the plug on something that is evolving, growing, accumulating knowledge all the time. It would be like stopping somebody's heart. When they're brought back, they're never quite right and you don't do it just because they've got a bit weirded out.'

'This isn't getting us anywhere. I fear neither of us is accumulating much knowledge. Polly has stalled Marvin, like an antiquated car, and either you are going to do this or I am.'

'You could damage her permanently,' Marvin said, fearing he might lose his sparring partner. Polly drove him mad much as a younger sister would. His frustration with her was his prerogative, but he also felt a duty to protect her.

'I'm asking you not to do this. If we wait, she'll come back.'

'We've waited long enough.'

DOI: 10.1057/9781137374851.0008

'But you don't even know how to do it.' Marvin sounded petulant now.

Janet started to move toward the main power supply, outside at the end of the corridor. Marvin had no choice but to call her back.

'Wait. At least let me do it properly. I'll close down her functions one at a time. It will take a few minutes. At least give me that.'

'Fine,' said Janet, returning to her desk. It took about twenty minutes. Marvin's face darkened then drained as Polly wound down. He really didn't know if they were going to get her back intact. She wasn't exactly the latest model and he couldn't help but think of the Emos as delicate. Janet was making him do this, but she couldn't make him witness the outcome. He restarted Polly's functions in the same order he'd shut them down, and when he finished he walked out and closed the door behind him.

Ignoring Marvin's dramatic exit, Janet refreshed her scheduler. By the time the name flashed up on her screen, she had already stood up and was watching what she thought was a joke, an annoying if clever hoax that Marvin, she assumed, must somehow have mocked up and planted in Polly. He was capable of it; that was obvious. He had the skills and, as he had let slip, the motivation. The little runt. He was either already in the pocket of Moogle Earthtechnics or he was planning to be and was prepared to use her as a means to an end. No wonder he didn't want to turn Polly off – and how convenient that he had left the room. Janet was watching herself preconsuming a ME Bio-home. She had touched the living walls in her living room, leaving handprints in the dense, luxury-grade grass. She had picked her own flowers, already individually potted in an interior hydroponic bed and was now demonstrating how to get varied lighting effects from a chandelier by feeding it different types of bacteria. Having done this, she exchanged a rather formal greeting with her Robocat and approached the kitchen-dining area, where she leaned against a transparent partition and dreamt of her future life in this future home. She was wearing a shimmering green and blue floral frock that evoked the biomaterials around her and, for Janet, the tinted photo she'd seen of her great grandmother back in the 1950s. There was a picture of her wearing a pencil line knee-length dress much like this. She was holding her infant daughter and standing at the back of her house. Janet always thought her expression was equivocal, as if she was unsure why or for whom she'd gotten dressed up. Janet saw herself seemingly more certain as, with her eyes closed and her head back, she exuded information. Some of this was obvious – her own vital statistics plus day, date, time, temperature and so on. All of it was displayed on

DOI: 10.1057/9781137374851.0008

the partition behind her. But as she continued to lean on this, it began to calculate, for example, her energy levels as a ratio of blood sugar, atmospheric conditions, rate of exertion and time expired since sleep. This was evidently not a quick sum, and while the numbers ran in one corner, words appeared and disappeared in the other. Words such as 'Shanghai', 'PFI', 'G30', 'suitcase', 'Shareesh', 'sixty eight', 'trap', 'stop', 'fuckers' formed a shape-shifting cloud of memories, feelings and thoughts, some of which Janet hadn't even had yet.

Unaware of what was going on behind her, onscreen Janet opened her eyes and stood away from the wall, now blank apart from the subtle-cerise ME logo and color-coded edging designed to prevent the occupant from walking into it. Off-screen Janet thought she saw herself addressed, for a moment, by a glance, a deliberative blink too uncertain and too fleeting to interpret but that created in her, before she had time to feel victimized, the sense of a self-conscious act that made every subsequent action, each demonstration appear hyperbolic. Aside from all the vegetation and other manifestations of bacteria-based lighting – including a family of green fluorescent bunny lamps made, wittily, from actual green fluorescent bunny cells – the living pod was a veritable palace of glass. Glass constituted each surface that wasn't green or glowing and part-constituted those that were. Although there were fewer screens than she was used to, everything that was flat or upright was in some way communicative. As if she'd snorted a bit of speed or taken deportment lessons from Marvin, she strutted into the kitchen and planted her hands on first the worktop then the fridge door then the wall, admiring the recipes, notices and news that emerged as if she had created them herself. Her solemn satisfaction rose again when she placed her biometric identification card on the dining room table and had it call one of her contacts by pointing at a 3D mug shot that was projected in miniature around the border of her own. Speech recognition was limited to certain keywords and phrases, but she was able to show how easy it was to order goods and services through a combination of touch and gesture that was more natural, intuitive and immediate than having to use a keyboard. Why bother to type or write when it was quicker to point, flick, flap, clap or stroke? Humans only learned to write because they had to, because the world was unresponsive to their needs and desires and they had to figure out how to make it so. Now everything was there, clear, open, transparent. Much of the time, Janet heard herself announce, we don't even need to ask! This technology will bring the answer to you! It knows

DOI: 10.1057/9781137374851.0008

what you want because it knows you. What it doesn't know, it is quite prepared to learn. The roles have been reversed. You no longer have to adapt to the technology because it will adapt to you. It will be as if it isn't there. Everything you want, as and when you want it: *Your world. Your way*. Want to view the entertainment chart? Just say 'menu', select media (at your super-sensor hotspots you can do this with your eyes!), pick a category (don't worry, we have them all), choose your medium and your tariff (remember, immersive mode means more thrills but sometimes also spills) and then leave it all to us. No need to pause, because your media can change rooms with you (yes, even there!). What about your husband (or partner!) and children? That's what the MEGoggles are for. You get your Romance, he gets his Sport, they get their Cartoons and everyone is happy. With our new Multichannel Role-player Parental Voice Simulator, we can even do the reading at bedtime so it never has to be your turn again! Need some groceries? All kitchen appliances and storage units have specialist databases and self-audit functions, meaning that they know what they have in stock and what they need to order. Obviously, they are networked to the central hub and have full access to your past and probable consumption habits. Tech support? Taken care of. Your ME Bio-home living space is fully automated and capable of self-diagnosis and repair. This service is included in your regular Habitation Package. Accidents and Outbursts are monitored and charged on a case-by-case basis using the universal insurance scale of moderate, neglectful and malevolent damage.

Since when had preconsumption become sales, and where was onscreen Janet's digital assistant? Had the roles merged, and in which case who or what was Janet looking at? Her onscreen self had been waiving her hands and arms around, making signs and pulling faces in a parody of mime or as though she was telling a story to a room full of kids. Janet watched as she made a patronizing pass by a little writing desk, rendered in wood to emphasize its antiquity. The only thing on the desk was a small photo frame performing a slide show of her life from infancy onwards. As if she did not care to see how far it would go, the Janet who was virtual moved on determinedly, running her hands over cool, sensuous material, using her body to allude to water and boats and lazy summer days while leaving behind her a turbulent trail of data.

DOI: 10.1057/9781137374851.0008

7
Conclusion: *i*Media Otherwise

Abstract: *Insofar as reality and subjectivity are always already mediated, what is at stake in the processes of unmediation and re-differentiation that have been traced in this book? Unmediation is more than the alignment of immediacy, masculinism and scientism in the quest to be right (versus writerly). It is about the presencing of iworlds just as they are – or very soon will be – and the moral and religious imperative to get real, man up, shape up, improve ourselves, perfect ourselves, save ourselves or else be damned! I confront the moralism and religiosity of unmediation, its right(eous)ness with a feminist tradition of writerliness and manifest mediacy that might allow space and time for the emergence of iworlds that are otherwise.*

Keywords: immediacy; mediation; unmediation

Kember, Sarah. *iMedia: The Gendering of Objects, Environments and Smart Materials.* Basingstoke: Palgrave Macmillan, 2016. DOI: 10.1057/9781137374851.0009.

A meditation on unmediation

In this book, I have identified and problematized a new dialectics of structure and scale, objects and relations and epistemology and ontology. More specifically, I have questioned the conceptual turn from structure to scale, epistemology to ontology and from subjects and objects to environments of processual and *i*mperceptible things-in-themselves. By asking who sees at the scale of imperceptibility (and from where – i.e., everywhere and nowhere), I have suggested that the old and unfashionable problem of masculine disembodied knowledge practices does not go away because we no longer deign to speak of it – because we do not do epistemology anymore. Indeed this problem resurfaces with a vengeance, propelled, as I've suggested it is, by some familiar allies. Allied to the new masculinism – one that is by no means exclusive to *i*media theory but that certainly characterizes it – is a renewed recourse to scientism and to immediacy manifested as unmediacy. My conclusion therefore begins with a meditation on unmediation in always already mediated environments within which, and about which we – always already mediated and re-differentiated subjects – might re-identify, following Xavière Gauthier, a 'frightful masculine fashion' of 'speaking in order to be right – how ridiculous!' (1985: 200). I have contrasted, or rather, put into tension, a masculinist tendency to rightness and a feminist tradition of writerliness. Writerliness is akin to hypermediacy, the manifest display of mediation that for Richard Grusin was once about the formal features of media – in our case, words, but words attended and attenuated by other signs, such as images, sounds and code – and is subsequently more about its infrastructure (2010). Either way (and for me, writerliness is itself multiscalar, incorporating inscription of many kinds at many levels), the manifest display of mediation through writerliness is at odds with the tendency toward rightness and the rediscovery of the real after the real – or after the division between the real and ideal, nature and language. Unlike Bergson, contemporary philosophers such as Harman and Hansen do not seek to trouble this division insofar as they consider themselves to have already transcended it (1992; 2010; 2013). The claim to be already post (meaning after) dialectical is among the conditions of possibility for the new dialecticism.

The new dialecticism is informed by physics envy. What Harman wants, for example, is to speak of wood itself, to grasp its material reality

DOI: 10.1057/9781137374851.0009

without recourse to representation. Unlike Barad, object oriented philosophers – OOPs theorists, as I have referred to them – are not interested in meeting the universe halfway (2007). Forming a bridge with OOPs theorists, as I've suggested they do, dialectical feminists such as Grosz and Bennett look to matter itself either as ground or foundation or in the spirit of affirmation (2005; 2001 and 2010). I understand scientism now to take the form of a turn from the humanities and social sciences, from literary and cultural theory and from sociology toward physics, biology and ecology. What is on offer here is not a conventional form of scientific realism but precisely the real after realism, after the division of real and ideal, nature and language. This notion of real is already post (meaning after) dialectical. It is always already mediated. This already mediated reality, the *i*world, as I have called it, is presenced through a process of unmediation, a process enabled, and indeed fuelled by, moralism and religiosity.

I have attended to the morality of glass, the *i*material of the day, the invisible maker up of *i*worlds. Glass is the very matter of unmediation: it presences the *i*world just as it is (or very soon will be), openly and transparently. Glass obviates the need for politics – because everything is already open and transparent – but conveys, without communicating, a clear morality, a moral imperative not just to be open and transparent but to man up, face up to reality, to what is, or soon will be. Google's glasses may be on hold, but the message they convey about the confluence of masculinity and unmediacy resonates in the prospect of twenty first century glassworlds, whether or not they ever fully scale up. Corning's glassworlds have, historically, had other subjects in mind. All the way from the cleanability of Pyrex to the flexibility of Willow Glass™, Corning glass has been less about manning up than shaping up, less about masculinity and more about the problem of femininity and its domiciles; the home (the kitchen), the body, the workplace. Technology has always compensated for perceived inadequacies in female efficiency, health, hygiene, productivity and reproductivity, but there is something else going on now, a fairy tale, a Cinderella story – if not about scullery maids becoming princesses, then about glass slippers becoming glass subjects. It is a case, as it always has been, of shaping up or getting shipped out. Glass itself, already plastic, promises to become elastic, like skin. Already hard at work, glass promises to transform itself into some thing organismic, smart, flexible, clean, slim, hard working, responsible

DOI: 10.1057/9781137374851.0009

and productive. Glass *is* feminized and optimized, the ultimate neoliberal subject.

The morality of unmediation – man up, shape up or ship out – is framed by religiosity, a familiar discourse of salvation and damnation, the hope of becoming better versions of ourselves, better citizens of the planet and the despair of being bettered by the consequences of failure. Current forms of religiosity have a distinctly ecological flavor. They are organized with respect to the anthropocene, the era of human history and of technological acceleration that is nearing an endpoint in the zenith of sustainability or the nadir of entropy and extinction. The humanism of the salvationists – and it is, I've argued, a hu-man-ism – is only lightly veiled in the damnationist's deafening cry: in my end is my beginning! I take as laughably disingenuous all forms of post – meaning after – humanism; all apparently logical exposition about the demise of human logics; all political treaties about the end of politics; all critical statements about abandoning critique and all writing about the end of writing. For me, critique is not abandoned but rather severely limited by crude substitutions that include the apparent belief that since critique and so much paranoia and negativity have failed to change anything or to deliver political alternatives, the solution is therefore to shift to where – i.e. in the natural, dynamic and ever-evolving world – change is. The affirmation of a dynamic, already mediated *i*world becomes a final ground for politics; the site of its vitality and continual re-emergence.

I have argued against the quest for final grounds and seek instead to establish conflicts, tensions and antagonisms. I invite those who embrace apocalypse and extinctionism, those who consider the world to be beyond the human entirely to address Hélène Cixous when, finding her selves stuck between a rock and a hard place, the abyss and the Medusa, she declared: 'Let's get out of here!' (1985: 255) A preference for generic subjects is part of what aligns the gods of theory – who see every imperceptible thing from nowhere – with the architects of *i*worlds. Their tactic of anonymity is a technique of power, one that is not extended to subjects who, as they always have, more than they ever have, find themselves the object of scrutiny, containment and moral judgment. Following Cixous, the gods and architects should beware of the configuration – potentially strategic – of goddesses, monsters and women. The laugh of the Medusa is still to be mimed.

DOI: 10.1057/9781137374851.0009

Is there a time machine for women?

The tension between being (trapped) and being beyond can be framed in terms of temporality and political subjecthood. In my chapter on ubiquitous women, I asked if there was a time machine for women. The question was not entirely facetious. There are clearly time machines for men, *histories* of heroic time travel that are relived, time and again, in fantasies of transcendence and omnipotence. There is no such magic for women, nothing that transports us back and forth, no machine that is subject to female feminist wills. If there were, we'd have got out of here by now. What is possible, I've speculated, is something quite provisional, a time machine predicated on the boundary work and boundary states of potential and *potentia*; structured, managed and optimized time and time that is indivisible, emergent and fundamentally unmanageable. I have refigured Microsoft's Ayla and Corning's Jennifer in the central character of my novel in progress. Ayla and Jennifer, like Cinderella are slavishly hard working (slim, flexible, efficient) and potentially perfect princesses. They meet and morph in Janet whose day, in *A Day in the Life of Janet Smart,* is quite a day. Janet's day regulates her in time while extending her as time, as life, as her desire to become. By living through it, to the extent that she lives through it, she becomes the agent of her own reconstitution as a political subject, a reconstitution that occurs in the tension between *homo oeconomicus* and its constitutive outside. As my own sf figure, Janet enacts an irony in the depressive, non-paranoid sense of holding contradictory things together – because both are necessary and true. She is at once a feminized neoliberal subject and its feminist, queer alternative. She is a figure of irony and parody understood here as unstable antagonisms, as openings to the political – more or less. Janet is my key to *i*media otherwise.

What can writing still do?

Following Cixous following Derrida, I maintain that the question of what writing can still do is a better question, a more effectively posed problem than the one concerning what writing is (1985; 1981). I remain interested in the non-identity, yet specificity of writing as literal

DOI: 10.1057/9781137374851.0009

inscription, including in contexts of contemporary art and feminist manifestos. But what I inherit is a sense of what writing (writing that was always bodily and technological, always post, where post does not mean after the human) can do under conditions of possibility and impossibility that include a marketized and privatized academy, a commercialized publishing industry and a more broadly neoliberalized culture (Kember 2015). As the principal mechanism of deconstruction, writing destroys order and creates openings to the political, albeit more or less. The instability and uncertainty of this opening is what is foregrounded in the use of parody and irony both of which can miss the mark, fail to create tension, reinforce as much as subvert sexism and misogyny. I advocate the use of these queer feminist writing strategies despite or because of their instability. My somewhat perverse position is that they might uncloak the cloak of humor and render absurd the new theatres of the absurd constituted by mishearing communication devices, networked mirrors, toasters and toilets, vibrating pants, exploding bras and the like. For me, what is yet to be mimed from the laugh of the Medusa is an antagonism in action, in the space and time of the laugh. This does not amount to a final ground for politics but might militate against its total, smart, systematized enclosure. Writing, including especially ironic and parodic writing, addresses the question that continues to arise in political theory, namely, what is to be done, what should we or what could we do? One answer has already been written: 'Write!'

I have written within and across forms that I consider to be distinct yet relational: the monograph, manifesto and novel. *iMedia* is a book that seeks to recognize the specificity and non-identity of writing and its genres. It picks up old and ever-new provisional forms such as the manifesto and refuses the dialectic of creative and conceptual, analogue and digital. It is a necessarily and im/possibly post-dialectical book presenting a necessarily as well as im/possibly post-dialectical argument. There is, as I've argued, no post-dialectical feminism. There is only the attempt to think of a dialecticity of dialectics that is itself not dialectical. I have suggested that the theorists and philosophers of *i*media, including feminist theorists and philosophers, are either trying too hard to do this, or not hard enough. The mistake, always, lies in thinking ourselves to be already post – meaning after – dialectical. So I cannot affirm or negate the existence of post-dialecticism but, once

DOI: 10.1057/9781137374851.0009

more, I do regard it as a strategy for creating tension within the otherwise totalizing scope of neoliberalism. The political opening created by such tension may never be certain, but it is made uncertainly available by foregoing final grounds and endless, ultimately delimiting substitutions.

DOI: 10.1057/9781137374851.0009

Bibliography

Adam, A. (1998) *Artificial Knowing. Gender and the Thinking Machine*, London and New York: Routledge.

Adams, T. (2014) 'The world of work is about to get a makeover. Meet your new bosses: data and algorithms', *Observer Tech Monthy*, 11.05.14.

Andrejevic, M. (2007) *iSpy. Surveillance and Power in the Interactive Era*, Lawrence: University Press of Kansas.

Andrejevic, M. (2013) *Infoglut. How Too Much Information Is Changing the Way We Think and Know*, New York and London: Routledge.

Armstrong, I. (2008) *Victorian Glassworlds. Glass Culture and the Imagination 1830–1880*, Oxford: Oxford University Press.

Barad, K. (2007) *Meeting the Universe Halfway. Quantum Physics and the Entanglement of Matter and Meaning*, Durham and London: Duke University Press.

Barthes, R. (1982) *Camera Lucida*, London: Flamingo.

Bates, L. (2014) *Everyday Sexism*, London: Simon and Schuster.

Beckett, S. (1958) *Endgame*, London: Faber and Faber.

Bennett, J. (2001) *The Enchantment of Modern Life. Attachments, Crossings, and Ethics*, Princeton and Oxford: Princeton University Press.

Bennett, J. (2010) *Vibrant Matter. A Political Ecology of Things*, Durham and London: Duke University Press.

Bergson, H. (1992) *The Creative Mind. An Introduction to Metaphysics*, New York: Citadel Press.

Birchall, C. (2011) 'Introduction to 'Secrecy and Transparency", *Theory, Culture & Society*, 28(7–8), pp. 7–25.

DOI: 10.1057/9781137374851.0010

Blaszczyk, R. L. (2000) *Imagining Consumers. Design and Innovation from Wedgwood to Corning*, Baltimore and London: The Johns Hopkins University Press.

Bogost, I. (2012) *Alien Phenomenology or What It's Like to Be a Thing*, Minneapolis, London: University of Minnesota Press.

Bolter, J. D. and Grusin, R. (2002) *Remediation. Understanding New Media*, Cambridge, Massachusetts and London, England: The MIT Press.

Braidotti, R. (2006) *Transpositions*. Cambridge: Polity Press.

Braidotti, R. (2010) 'Powers of affirmation: Response to Lisa Baraitser, Patrick Hanafin and Clare Hemmings', *Subjectivity*, 3(2), July 2010, pp. 140–149.

Brown, W. (2005) *Edgework. Critical Essays on Knowledge and Politics*, Princeton and Oxford: Princeton University Press.

Brownell, B. (ed) (2010) *Transmaterial 3. A Catalog of Materials That Redefine Our Physical Environment*, New York: Princeton Architectural Press.

Brownell, B. (2012) *Material Strategies. Innovative Applications in Architecture*, New York: Princeton Architectural Press.

Bruining, D. (2013) 'A Somatechnics of Moralism: New Materialism or Material Foundationalism', *Somatechics* 3(1), pp. 149–168.

Brynjolfsson, E. and McAfee, A. (2014) *The Second Machine Age. Work, Progress, and Prosperity in a Time of Brilliant Technologies*, New York and London: W. W. Norton & Company.

Carter, A. (2008) *Little Red Riding Hood, Cinderella, and Other Classic Fairy Tales of Charles Perrault*, London and New York: Penguin Books.

Christian, D. (2011) *Maps of Time. An Introduction to Big History*, Berkeley, Los Angeles, London: University of California Press.

Cixous H. (1985) 'The Laugh of the Medusa' in E. Marks and I. de Courtivron (eds) *New French Feminisms*, Sussex: The Harvester Press Limited, pp. 245–264.

Clarke, T. (2012) 'Scale', in T. Cohen (ed) *Telemorphosis. Theory in the Era of Climate Change*, Ann Arbor: Open Humanities Press, pp. 148–167.

Cohen, T. (ed) (2012) *Telemorphosis. Theory in the Era of Climate Change*, Ann Arbor: Open Humanities Press.

Colebrook, C. (2004) *Irony*, New York and London: Routledge.

Cruikshank, G. (1854) *Cinderella and the Glass Slipper*, London: D. Bogue.

Dean, J. (2009) *Democracy and Other Neoliberal Fantasies. Communicative Capitalism and Left Politics*, Durham & London: Duke University Press.

DOI: 10.1057/9781137374851.0010

Dentith, S. (2000) *Parody*, New York and London: Routledge.

Derrida, J. (1981) *Writing and Difference*, London: Routledge and Kegan Paul Ltd.

Derrida, J. (1997) *Of Grammatology*, Baltimore: John Hopkins University Press.

Derrida, J. (2001) 'I Have a Taste for the Secret', in J. Derrida and M. Ferraris *A Taste for the Secret*, Cambridge: Polity Press.

Erickson, K.; Kretschmer, M. and Mendis, D. (2012) *Copyright and the Economic Effects of Parody*, Bournemouth University: Centre for Intellectual Property Policy and Management.

Friedan, B. (2010) *The Feminine Mystique*, London: Penguin Books.

Fuhrt, B. (ed) (2011) *Handbook of Augmented Reality*, New York, London: Springer.

Gane, N. (2006) 'When We Have Never Been Human, What Is to Be Done? Interview with Donna Haraway', *Theory, Culture & Society*, 23(7–8), pp. 135–158.

Gauthier, X. (1985) 'Why witches?' in E. Marks and I. de Courtivron (eds) *New French Feminisms*, Sussex: The Harvester Press Limited.

Gill, R. and Pratt, A. (2008) 'In the Social Factory? Immaterial Labour, Precariousness and Cultural Work', *Theory, Culture & Society*, 25(7–8), pp. 1–30.

Gill, R. (2010) 'Breaking the silence: the hidden injuries of the neoliberal academy', in R. Ryan-Flood and R. Gill (eds) *Secrecy and Silence in the Research Process: feminist reflections*, London and New York: Routledge, pp. 228–245.

Gill, R. and Scharff, C. (2013) *New Femininities: Postfeminism, Neoliberalism and Subjectivity*, Hampshire and New York: Palgrave Macmillan.

Greenfield, A. (2006) *Everyware: The Dawning Age of Ubiquitous Computing*, Berkeley, CA: New Riders Publishing.

Grosz, E. (2005) *Time Travels. Feminism, Nature, Power*, Durham and London: Duke University Press.

Grusin, R. (2010) *Premediation. Affect and Materiality after 9/11*, Hampshire and New York: Palgrave Macmillan.

Hall, G. (2012) 'White Noise: On the Limits of Openness (Living Book Mix)', *Digitize Me, Visualize Me, Search Me*, www.livingbooksaboutlife.org

Hansen, M. B. (2013) 'Ubiquitous Sensation: Toward an Atmospheric, Collective, and Microtemporal Model of Media' in U. Ekman (ed)

DOI: 10.1057/9781137374851.0010

Throughout. Art and Culture Emerging with Ubiquitous Computing, Cambridge Mass: The MIT Press.

Haraway, Donna, J. (1991) *Simians, Cyborgs and Women. The Reinvention of Nature,* London: Free Association Books.

Haraway, Donna, J. (1997) *Modest_Witness@Second_Millennium,* New York, London: Routledge.

Haraway, D. (2011) *SF: Speculative Fabulation and String Figures,* Kassel: dOCUMENTA (13).

Harman, G. (2010) *Towards Speculative Realism,* Hants: Zero Books.

Hayles, N. Katherine (1991) *Chaos and Order. Complex Dynamics in Literature and Science,* Chicago and London: University of Chicago Press.

Hayles, N. Katherine (1999) *How We Became Posthuman. Virtual Bodies in Cybernetics, Literature and Informatics,* Chicago and London: University of Chicago Press.

Hemmings, C. (2014) 'The materials of reparation', *Feminist Theory,* 15(1), pp. 27–31.

Hutcheon, L. (1989) *The Politics of Postmodernism,* London and New York: Routledge.

Ingold, T. (2010) 'Bringing Things to Life: Creative Entanglements in a World of Materials', *Realities,* Working Papers No. 15, ESRC National Centre for Research Methods, www.ncrm.ac.uk

Ingold, T. (2011) *Being Alive. Essays on Movement, Knowledge and Description,* London and New York: Routledge.

Jameson, F. (1991) *Postmodernism, Or, the Cultural Logic of Late Capitalism,* Durham and London: Duke University Press.

Kember, S. (2003) *Cyberfeminism and Artificial Life,* London and New York: Routledge.

Kember, S. and Zylinska, J. (2012) *Life After New Media. Mediation as a Vital Process,* Cambridge, Massachusetts and London, England: The MIT Press.

Kember, S. (2013) 'Ambient Intelligent Photography' in M. Lister (ed) *The Photographic Image in Digital Culture,* second edition, London and New York: Routledge, pp. 56–77.

Kember, S. (2014) 'Why Write? Feminism, Publishing and the Politics of Communication', *New Formations* Number 83, pp. 99–117.

Klein, M. (1988) *Envy and Gratitude and other works 1946–1963,* London: Virago.

DOI: 10.1057/9781137374851.0010

Lindström, K. and Ståhl, Å (2014) *Patchworking Publics-In-The-Making. Design, Media and Public Engagement*, Malmö: Malmö University.

Lyon, J. (1999) *Manifestoes. Provocations of the Modern*, Ithaca and London: Cornell University Press.

Lyotard, J. F. (1984) *The Postmodern Condition: A Report on Knowledge*, Manchester: Manchester University Press.

Manzini, E. (1989) *The Material of Invention. Materials and Design*, London: The Design Council.

McNeil, M. (2010) 'Post-Millennial Feminist Theory: Encounters with Humanism, Materialism, Critique, Nature, Biology and Darwin', *Journal for Cultural Research*, 14(4), pp. 427–437.

McRobbie, A. (2010) *The Aftermath of Feminism. Gender, Culture and Social Change*, London, New Delhi: Sage.

Miodownik, M. (2013) *Stuff Matters. The Strange Stories of the Marvellous Materials That Shape Our Man-Made World*, London: Viking.

Mouffe, C. (2005) *On the Political*, London and New York: Routledge.

Mouffe, C. (2009) *The Democratic Paradox*, London and New York: Verso.

Mouffe, C. (2013) *Agonistics. Thinking the World Politically*, London and New York: Verso.

Nakashima, H.; Aghajan, H. and Augusto, J.C. (eds) (2010) *Handbook of Ambient Intelligence and Smart Environments*, New York, London: Springer.

Perec, G. (2003) *Life. A User's Manual*, London: Vintage.

Perrault, C. (1697) 'Cendrillon, ou la petite pantoufle de verre', *Histoires ou Contes du temps passé, avec des moralités*, Paris.

Phillips, J. W. P. (2011) 'Secrecy and Transparency. An Interview with Samuel Weber', *Theory, Culture & Society*, 28(7–8), pp. 158–172.

Prigogine, I. and Stengers, I. (1989) *Order Out of Chaos*, Dell Publishing Group: Bantam Doubleday.

Reynolds, J. (2010) 'Jacques Derrida (1930–2004)', Internet Encyclopedia of Philosophy http://www.iep.utm.edu/derrida/

Rossiter, N. (2015) 'Coded Vanilla: Logistical Media and the Determination of Action', *South Atlantic Quarterly* 114(1), pp. 135–152.

Schmidt, E. and Cohen, J. (2013) The New Digital Age. Reshaping the Future of People, Nations and Business, London: John Murray.

Seghal, M. (2014) 'Diffractive Propositions: Reading Alfred North Whitehead with Donna Haraway and Karen Barad', *parallax*, 20(3), pp. 188–201.

DOI: 10.1057/9781137374851.0010

Shteyngart, G. (2011) *Super Sad True Love Story*, London: Granta.

Sobchack, V. (1990) 'A Theory of Everything: Meditations on Total Chaos', *Artforum*, 29(2), pp. 148–155.

Suchman, L. (2007) *Human-Machine Reconfigurations*, Cambridge: Cambridge University Press.

Warner, M. (1995) *From the Beast to the Blonde. On Fairy Tales and Their Tellers*, London: Vintage.

Weber, S. (2000) 'The Future of the Humanities: Experimenting', *Culture Machine,* Vol. 2 http://www.culturemachine.net/index.php/cm/article/view/311/296

Wiegman, R. (2014) 'The times we're in: Queer feminist criticism and the reparative 'turn'', *Feminist Theory*, 15(1), April 2014, pp. 4–27.

DOI: 10.1057/9781137374851.0010

Index

DOI: 10.1057/9781137374851.0011

DOI: 10.1057/9781137374851.0011

DOI: 10.1057/9781137374851.0011

Lightning Source UK Ltd.
Milton Keynes UK
UKOW01n1216240316

270816UK00003B/7/P